LD

visited by
ther floras partially
tion (Grey)

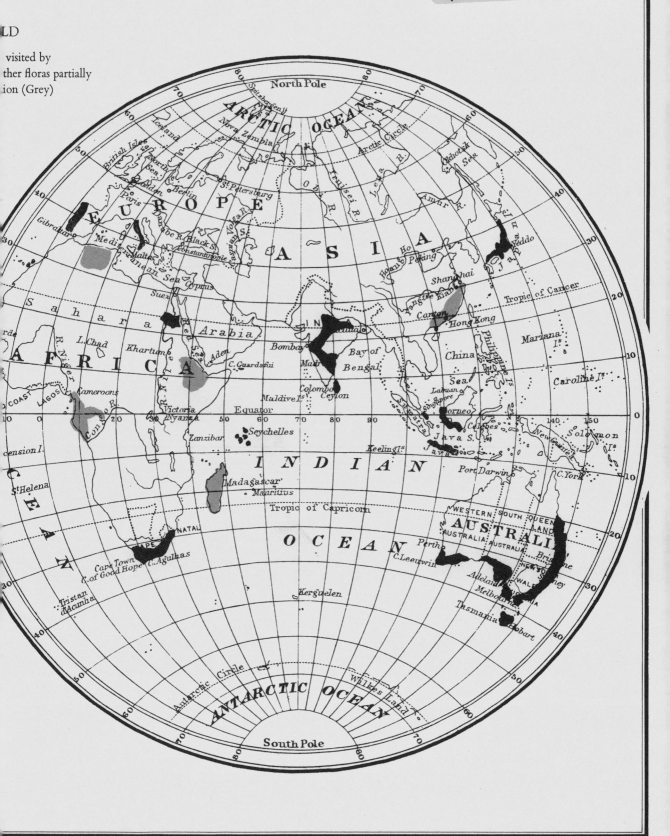

A
Vision of
Eden

Marianne North
Painted by Mary Hall, at Hastings Lodge, Hastings, 1866

A Vision of Eden

The Life and Work of
MARIANNE NORTH

Preface by Professor J. P. M. Brenan, M.A., B.Sc., F.C.S., F.I.Biol.
Former Director of the Royal Botanic Gardens, Kew

Foreword by Anthony Huxley

Biographical Note by Brenda E. Moon

In association with The Royal Botanic Gardens, Kew

LONDON: HMSO

© The Royal Botanic Gardens, Kew 1980
Applications for reproduction should be made to HMSO

First published 1980
Fourth impression 1993

British Library Cataloguing in Publication Data
A CIP catalogue record for this book is available from the British
Library

Reprinted for The Royal Botanic Gardens, Kew

Dd 293673 C40 3/93

Contents

Preface 7

Introduction 9

Publisher's Note 15

I. Early Days and Home Life 17

II. Canada and the United States 31

III. Jamaica 47

IV. Brazil 59

V. Teneriffe – California – Japan – Singapore 81

VI. Borneo and Java 95

VII. Ceylon and Home 115

VIII. India 123

IX. Second Visit to Borneo – Australia 151

X. New Zealand and the United States 183

XI. South Africa 199

XII. Seychelles Islands 217

XIII. Chili 225

XIV. Final Days 232

Biographical Note 234

List of Plants 240

As well as her outstanding contribution to art in natural history, Marianne North was also responsible for collecting a number of plants. Several species and one genus were later named after her.

Above left: Crinum northianum from Borneo. Although common in that country this species had not been previously described before collection by the artist.

Above right: A pitcher plant from the limestone mountains of Sarawak, Borneo (*Nepenthes northiana*).

Below left: Foliage, flowers and fruit of the Capucin Tree of the Seychelles (*Northea seychellana*).

Below right: A Giant Kniphofia (*Kniphofia northiae*) near Grahamstown, South Africa.

Preface

Professor J. P. M. Brenan, M.A., B.Sc., F.C.S., F.I.Biol., Former Director of the Royal Botanic Gardens, Kew

It was the *Pall Mall Gazette* which first suggested that Marianne North's pictures should come to Kew, when it reviewed an exhibition she held in 1879. Miss North was looking for a suitable home for her paintings as they made her Victoria Street flat more than a little crowded, and she wrote to my predecessor Sir Joseph Hooker asking if he would agree to the building of a gallery for them in the Gardens, at her own expense. Needless to say, Sir Joseph was happy to accept her offer.

Miss North arranged for her friend James Fergusson, a well-known architectural historian, to design the building, and it has been noted that the Gallery resembles a Greek temple with oriental verandahs. This was due partly to the architect's extensive knowledge of Indian architecture but also to his strongly held views on the way in which Greek temples had been lit; hence the clerestory windows, a departure from the established methods of lighting galleries by means of skylights. They were quite successful, except in dull weather, but artificial lighting was not installed until some twenty years ago, and that caused unfavourable comment from the purists.

Marianne North arranged the pictures, and the way in which this was done, close together with no wall showing, has proved a source of difficulty to later generations. When a picture has to be removed for restoration or photography others have to be removed from the top downwards in order to get the one wanted; ten pictures may have to be moved to get one! In spite of this, suggestions that the pictures be rearranged so that only a selection is on show at a time have always

been turned down. Reasonably enough, I think, as a considerable part of the visual impact of the Gallery is due to the way in which the eyes are assaulted by the great mass of colour on entering.

Miss North's generosity did not end with the building of the Gallery and its ancillary rooms. She also arranged for a catalogue to be compiled by the botanist W. Botting Hemsley who had been on the staff at Kew, but had retired because of bad health. Fortunately this later improved and from 1890 to 1908 he was Keeper of the Herbarium and Library here. He was a careful, scholarly man as those who have seen copies of the old catalogue (long out of print) will recognise. The first edition of 2000 copies, financed by Miss North, sold out some six months after the Gallery opened in 1882. At her request, the proceeds were used to buy books for the young gardeners' library, which was used in the evenings after work. Subsequent editions were undertaken by the Stationery Office.

The position of the Gallery was chosen by Miss North for a number of reasons. She felt, correctly, that a shelter and resting place was needed in that part of the Gardens for those who had toured the glasshouses; she also hoped that at some time there would be an entrance gate near the Gallery, and the site she suggested is indicated by the short stretch of railings along the Kew Road between the Gallery and the Temperate House Lodge. The gates never materialised and that part of the Gardens remains quieter than the area between the Main Gates and the Palm House where, as Miss North said, visitors "promenaded." She did not expect them to find their way to the Gallery but a great many people did, although over the years the numbers dropped. In fact, from the 1920s onwards there have been articles in the press at intervals about this "hidden treasure" at Kew. Recently there has been a resurgence of interest in Marianne North and her paintings and I am therefore extremely glad to have the opportunity of welcoming this book and to say how glad we are that people all over the world are to have the chance of learning something about a very charming person and of seeing reproductions of some of her paintings. I feel sure this book will encourage a great many to come and see the pictures for themselves.

Introduction

Anthony Huxley

In 1882, when Marianne North's paintings were made accessible to the public at Kew Gardens, photography was in its infancy and television not even a Jules Verne dream. The British public, more and more aware of the diversity of life on earth, were avid for natural wonders and accounts of strange peoples. There were plenty of botanical explorers but they had to set down their experiences in writing, with the help only of engravings made from sketches: there was no way of achieving the instant access to audiences of millions which is the prerogative of the modern botanical traveller. Nor was it easy to travel in many countries: after long sea voyages it was a matter of horse-drawn transport, of riding, often of walking, along poorly made roads and tracks. Hotels were few and doctors scarce if illness struck.

It is against this background, at a time when women travelling on their own were rare and remarkable creatures, that Marianne North's achievement, and the thronging of her Gallery in the years succeeding its opening, should first of all be considered.

Skill with the paintbrush was not in fact unusual in young ladies of her time. It was but a year after her birth that J. C. Loudon wrote in his *Gardener's Magazine* that "to be able to draw flowers botanically, and fruit horticulturally . . . is one of the most useful accomplishments of your ladies of leisure, living in the country." A great many young ladies of leisure, indeed, had all too little to amuse themselves until – unless – they married. They were often taught music, as was Marianne,

as well as painting, and the combination of painting and an interest in plants was frequently strong.

If Marianne North had married – if she had not had a widowed father whom she looked after and travelled with until his death – if she had not subsequently had adequate means – her works might still be largely unknown, like the flower paintings of her sister Catherine who married John Addington Symonds.

Miss North, however, was not only reasonably talented but fortunate. She travelled quite widely with her father in her earlier years, sketching and painting as they went. On his death, when she was almost 40, she overcame her grief by painting, and later set her sights on recording tropical crop plants and flowers. The urge to travel deeply ingrained, the possibility of marriage apparently never offered, combined to make her continue her wanderings and painting for the next fifteen years.

In her youth, at the same time as acquiring dexterity with the paintbrush, she must also have learned a good deal of botany. In the first place she was early on an enthusiastic gardener who cultivated both hardy and greenhouse plants. She recorded that when in London she often visited the Royal Horticultural Society's Chiswick gardens (as they were then) and the Royal Botanic Gardens at Kew; and it is clear that while in her twenties she became acquainted, presumably through her father, with the Director of Kew, Sir William Hooker. Later his son Joseph Hooker became Director and again it seems that he and Marianne were on good terms. She also met

The Marianne North Gallery at Kew Gardens, which houses 832
botanical paintings. The Gallery is situated south of the Victoria
Gate opposite the Temperate House and is open to the public all the
year round. It was designed under Miss North's supervision by
James Fergusson, the architectural historian, to a plan which
included features of Greek temple architecture, such as the

clerestory windows. Marianne North herself arranged the paintings and designed and painted the frieze and decorations surrounding the doors. These views show clearly the arrangement of the two inner rooms and the surround of 246 different types of wood which she collected on her travels.

A Vision of Eden

Charles Darwin who was in fact responsible for her deciding to visit Australia.

A painter and, close second, a really determined traveller – one is reminded of Lady Hester Stanhope in the Middle East or Mary Kingsley in Africa – Marianne North was also a fluent and prolific letter writer, and this ability she later developed to provide us, from letters, diaries and notes, with the free-flowing, carefree narrative of which this present book is an abridgement. I use the word carefree advisedly: there is practically no hint anywhere of the immense amount of planning her journeys must have involved, while scorching sun, drenching rain, fearful road conditions, travel sickness, leeches and giant spiders, and unsalubrious accommodation are all dismissed in a few airy words. She preferred a tent or barn to a formal drawing room, solitude to the company of local bigwigs, unpretentious people to servants of any kind. One must comment that the world she travelled in was one where strangers were often made incredibly welcome by today's standards, though she mentions surly compatriots as when, on one ship, she "was put among a mixed lot of Britishers, and never spoke a word for four days."

Difficulties of language and the vagaries of locals are hardly mentioned, though our author records at least one example of getting into "one of those rages which are sometimes necessary." It is only of cold places that we hear the occasional complaint, because she felt the cold badly and it brought on rheumatic pains which, if protracted, rendered her almost immobile as the years went by. Thus she speaks unhappily of New Zealand because it was so cold and wet when she visited it; yet her paintings from there are as good as any others and some of the views more than usually fine.

The first impression of the Marianne North Gallery is almost overpowering. Apart from the narrow picture frames, the walls are entirely covered with paintings, fitted together like a jigsaw or, as Wilfrid Blunt describes them, "like a gigantic botanical postage-stamp album ..." where "yet further flowers scramble up the door-posts and across the lintels." It is easy to take a rather cursory view and reel away bemused by sheer quantity, overall colour and a slightly appalled sense of incredible diligence, but a methodical, unhurried approach provides many rewards.

The paintings vary in size from a few square inches, often narrow horizontal oblongs, to the maximum of perhaps 40 by 15 inches devoted to a life-size rendering of the flowerhead of the great Chilean *Puya whitei* with its hundreds of florets. Apart from flowers, birds, insects and animals feature in many of the paintings; there are views of scenery with the emphasis on vegetation, some of which are pure landscapes while others show native buildings and local people. This varied approach must have been specially popular in the early years of the Gallery when, as I have suggested, people could glean so little visually about the topography of distant lands.

There is an equally varied approach to the plants which form the majority of the subjects. Sometimes we see the whole plant growing naturally in its setting; sometimes details of a flowerhead, of fruits, even of aerial roots in one case. I am not sure that the studies of fruits, usually with one peeled or pulled apart, are not the most pleasing of all. Many of the paintings are of useful plants; an unusual one shows the way in which prickly pears are used to nurture cochineal insects, being swathed in rags on which the insects have been hatched.

Some of the scenic views are partly framed by flowers close at hand – contrived pictures perhaps, but often very successful. Then there are many examples of flowers grouped: as her narrative explains there were many occasions when her local friends brought in quantities of different plants which she was always anxious to record. Again, the treatment of these varies. Some groups are arranged in containers, often interesting local ware like a maté gourd in Brazil: these might be likened to Dutch flower paintings in their treatment. Other groups appear laid on a table; in yet others, the plants are shown more or less as if growing, though tight-packed. One might say that Marianne North used her brush as the modern botanical traveller uses a camera, but in these groupings and some of the "contrived" scenes she achieves effects which a camera never could.

Another striking aspect of these paintings is the sheer variety of the plants portrayed. There are 832 paintings in the Gallery, and another 16 originally in the lobby now elsewhere. Although many of these are views, in total no less than 727 genera and approaching 1000 species are portrayed. Of course this is a small fraction of the potential total from the countries she visited, but nonetheless an excellent and varied sample in each case. (The Indian paintings, incidentally, include 28 of plants often mentioned in Indian literature, especially in connection with their religious significance.)

Many of the plants were barely known either botanically or horticulturally when she painted them, and four were previously unknown to science and named after her. One of these, *Crinum northianum* from Borneo, was actually described from her drawings. As the original catalogue says, this is "the highest compliment which could be paid to their scientific accuracy." And this leads me to a final point about these remarkable paintings – their instant recognisability. I am not really qualified to write of them as works of art, though I have suggested that the variety of approach is wider than many botanical artists achieve. In *The Art of Botanical Illustration* (1950) Wilfrid Blunt, judging them as art, criticised them as "almost wholly lacking in sensibility" and making "a disagreeable impression"; however, in his 1978 account of Kew, *In for a Penny*, he has mellowed towards them somewhat. Looked at closely, one can see what he meant by "the curiously dry and unattractive quality of the paint," but at a slight distance life truly emanates from the brushwork in a remarkable way, and one agrees more with the reporter who wrote up the Gallery's opening in the *Gardeners' Chronicle* for 6th June 1882, of the "skilful, dashing works of a true artist." I write here of the depictions of plants and vegetation more than those of native life, which are certainly sometimes a little stilted in their approach.

It is not seeking to excuse any failing in technique if one recalls that most of these paintings were carried out with minimum facilities, sometimes in a single day, and were often packed while still wet as Marianne continued her journeyings, forever impelled to move on. They were painted on prepared paper and fixed to canvas in England, when without doubt the artist went over them to conceal blemishes and damage due to travel. One might finally add that, apart from admitting to "oil-painting being a vice like dram-drinking," she never writes of technique or painting prob-

lems: once acquired, her art was a gift she took for granted. As a matter of interest, her technique shows little change throughout her period of travelling, although some of the Australian views are more impressionist than most, perhaps done in a hurry.

Most of the time Marianne found nature relatively unspoiled, as near Rio de Janeiro where "every rock bore a botanical collection fit to furnish any hot-house in England." But there are a number of references to the regrettable ravages caused by man: for instance of the Californian redwoods, "it broke one's heart to think of man, the civiliser, wasting treasures in a few years to which savages and animals had done no harm for centuries." In his Preface to the original catalogue Sir Joseph Hooker wrote too of how many of the habitats "are already disappearing or are doomed shortly to disappear before the axe and the forest fires, the plough and the flock, of the ever advancing settler or colonist. Such scenes can never be renewed by nature, nor when once effaced can they be pictured to the mind's eye, except by records such as this ..." Especially for these reasons one can, more than ever today, again echo the *Gardeners' Chronicle* reporter's comment that "such an adjunct to a botanical garden is unique."

Let us finally remember that not for nothing did Marianne North entitle the tale of her life and travels *Recollections of a Happy Life* (the third volume of her three-part autobiography, entitled *Further Recollections of a Happy Life*, contains material which was edited out of volumes I and II, and collated after her death by her sister Catherine Symonds). She must indeed have had a sunny disposition to surmount the frequent travails of her journeys and to receive such welcome from so many strangers. Let me conclude by quoting Edward Lear, who recorded how a pilot who had taken Marianne and her father up the Nile described her, at the age of about 25 (at that period the Arabic word *bint* had no derogatory connotations): "This Bint was unlike most other English Bints, being, firstly, white and lively; secondly, she was gracious in her manner, and of kind disposition; thirdly, she attended continually to her father ...; fourthly, she represented all things on paper ... she was a valuable and remarkable Bint!"

A Friendly Audience. From a sketch by R. Phené Spiers. This drawing shows
Marianne North sketching on Elephantine Island in January 1868.

Publisher's Note

The names used for countries, towns and cities in this abridged version of Marianne North's books are the same as in the original, and modern place names in the captions have been adjusted to comply with the main text. Many such names have either been completely changed in the past century or are now spelt differently. The following list shows the major changes that will be useful in assisting the reader. The original names appear first:

Aboo Simbel = Abu Simbel
Abyssinia = Ethiopia
Acrae = Palazzolo Acreide
Amberawa = Ambarawa
Assooān = Aswān
Aurungabad = Aurangabad
Batavia = Djakarta
Bendoeng = Bandung
Beyroot = Beirut
Bhartpur = Bharatpur
Boro-Bodo = Borobudur
Buitenzorg = Bogor
Cattaro = Kotor
Ceylon = Sri Lanka
Chili = Chile
Clifton = Niagara Falls City (Canada)
Constantinople = Istanbul
Cookstown = Cooktown
Dindigal = Dindigul
Djocia = Jogjakarta
Fiume = Rijeka
Frankfort = Frankfurt
Garoet = Garut
Girgenti = Agrigento
Jako = Jakko
Jamaica Plain = part of Boston
Java = Djawa, part of Indonesia

Kamaun = Kumaun
Madura = Madurai
Mentone = Menton
Nasirabad = Mymensingh
        (Bangla Desh)
Pasoeroean = Pasuruan
Pesth = part of Budapest
Pola = Pula or Pulj
Prambanan = Brambanan
Samarang = Semarang
San Gabrielle = San Gabriel
Sarawak = state of Malaysia
Sindang Sari = Sindanglaja
Smyrna = Izmir
Soerabaja = Surabaja
Soerakarte = Surakarta
Spalatro = Split
Spezzia = La Spezia
St. Gothard = St. Gotthard
Syra = Syros
Tanjore = Thanjavur
Temanggoeng = Temanggung
Teneriffe = Tenerife
Thibet = Tibet
Trichinopoli = Tiruchchirappalli
Winnepiseogee, Lake = Lake
        Winnipesaukee

T. Parkin
1893

CHAPTER I

Early Days and Home Life

1830–70

IT BEGAN AT HASTINGS IN 1830, but as I have no recollections of that time, the gap of unreason shall be filled with a short account of my progenitors. My fourth great grandfather was Roger, the youngest son of Dudley, fourth Lord North of Kirtling, and Anne, daughter of Sir Charles Montagu. He had been Attorney-General under James II, and wrote the lives of his three brothers – the Lord Keeper Guilford; Sir Dudley, Commissioner of the Treasury to King Charles the Second; and Doctor John North, Master of Trinity College, Cambridge. The portraits of these famous brothers and of their grandfather, the third Lord North, were among the first things which impressed me with childish awe, in our dining-room at home.

For Roger I had an especial respect, as the brown curly wig was said to be all his own, and not stuck on with pins driven into his head as my doll's wig was, and I thought he used to look down on me individually with a calm expression of approval. When he tired of being a lawyer and the political squabbles of the time, he retired to the old hall at Rougham, in Norfolk where he lived ("out of the way," he called it truly) to a good old age.

His son Roger had a vile temper, and flogged his son Fountain to such a degree that the boy ran away to sea, and stayed there till his father's death left him Squire of Rougham – a place he hated from old associations – so he never went near it again, and ordered the house to be blown up with gunpowder, as it was too solidly built to be pulled down easily. The sailor-squire cared only for the sea, and in his old age settled himself as close to it as he could in the first lodging-house ever let at Hastings, dividing his time between it and a house he built at Hampstead.

My grandfather Frederick Francis also lived all his life at Hastings, and never went near Rougham. He had five sons and a daughter, of whom my father was the eldest; he was born in 1800, and when a mere child of eight years old was sent to Harrow to fight his way among his elders, and endure many a hard hour of bullying and fagging. From Harrow my father went to St. John's, Cambridge, and in due time took his degree of Senior Op., spending his vacations with an old farmer at Rougham in preference to his ungenial home, and getting a liking for the old place, its noble trees and poor neglected people – a liking which increased with years. After leaving college he went to Switzerland, put himself on board with a Geneva family to learn French, walked round Mont Blanc, picked up some crystals, and finally came back to the Temple to study law, but instead of doing so, fell in love with my mother, the beautiful widow of Robert Shuttleworth of Gawthorpe Hall, Lancashire, and eldest daughter of Sir John Marjoribanks, Bart., of Lees, M.P. for Berwickshire. My mother's first marriage had been soon over; her husband having

The house at Hastings where Marianne North spent her childhood.

17

upset a coach and four he was driving, died himself and nearly caused the death of his wife and of the delicate child Janet, who was born afterwards. My father was elected member for Hastings in 1830 by ten "Freemen," one of them being himself.

My first recollections relate to my father. He was from first to last the one idol and friend of my life, and apart from him I had little pleasure and no secrets. I have a strong recollection of seeing the great dinner given after the passing of the Reform Bill, for which my father voted, riding or walking home night after night after the heated divisions to his house at Notting Hill, and arriving in the small hours of the morning. When that was over, his health broke down, and he had to give up Parliament for a while, and had the more leisure to attend to me. We had much variety in our life, spending the winter at Hastings, the spring in London, and dividing the summers between my half-sister's old hall in Lancashire and a farm-house at Rougham.

Governesses hardly interfered with me in those days. Walter Scott or Shakespeare gave me their versions of history, and Robinson Crusoe and some other old books my ideas of geography. The farm-house we lived in at Rougham had been originally the laundry of the Hall, and consisted of one large centre room on the ground-floor, with sufficient bedrooms over it and offices outside. The garden was full of old-fashioned flowers; it had tiny paths and beds edged with box hedges, leading up to a quaint old pigeon-house covered with ivy, and beyond that was the park full of grand trees, and the church and village. Our life at Hastings was very different, and our comfortable house was generally full of guests. My half-sister Janet had been a good deal away for some time with her cousin, Mrs. Davenport, afterwards Lady Hatherton, and one morning wrote the astonishing intelligence that she was engaged to be married to Dr. Kay, the great educationalist. Dr. Kay came down to see his future mother-in-law. He was twelve years older than Janet and very bald, and as he took my sister Catherine on his knee and petted her to hide his nervousness, she

made the deliberate and somewhat embarrassing remark, "Dr. Kay, why does your head come through your hair?" Terrible innocent! Catherine was about four years old then, I twelve, and Charley two years older. The wedding was all fun to us.

Our journeys from Hastings to Norfolk every year were a long week's work, and we were treated like old friends at all the inns on the road. My father often drove himself, with me on the box beside him. We also rode some hours each day. I knew every big tree, pretty garden, or old farm-house, with the wooden patterns let into the walls and yews and box-trees cut into cocks and hens, and I sadly missed them when the days of "improvement and restoration" came.

In August 1847 we went to Heidelberg, where we settled for eight months in the two upper storeys of a large ugly house outside the town gates, on the Mannheim road. Before April was over we left Heidelberg and spent the next two years touring Europe, often being too close for comfort to the civil disturbances and wars rife in Europe at the time.

In London during 1850 I had some lessons in flower-painting from a Dutch lady, Miss van Fowinkel, from whom I got the few ideas I possess of arrangement of colour and of grouping, and then we recommenced the happy old life at Rougham, I passing hours and hours of every day on horseback, painting and singing with little fear of interruption. The next season I saw the opening of the first great Exhibition (1851). Bartholomew gave me a few lessons in water-colour flower-painting; the only master I longed for would not teach, i.e. old William Hunt, whose work will live for ever, as it is absolutely true to nature. We used to see a good deal of him at Hastings, where he generally passed his winters, living in a small house almost on the beach under the East Cliff, where he made most delicious little pencil-sketches of boats and fishermen.

In May 1854 my father became again M.P. for Hastings, being elected without opposition on the death of Mr. Brisco.

On the 17th of January 1855 my mother died. Her end had come gradually; for many weeks we felt it was coming. She did not suffer, but enjoyed nothing, and her life was a dreary one. She made me promise never to leave my father, and did not

Opposite: The artist Marianne North at her easel.

18

A Vision of Eden

like any one to move her but him; he was always gentle and ready to help her, and missed her much when she was gone, writing in his diary in his own quaint way: "The leader is cut off from the main trunk of our home, no branches, no summer shoots can take its place, and I feel myself just an old pollard-tree." My father let Hastings Lodge, and took a flat in Victoria Street. Soon it became more like home than any other to me, and was a great rest after the big house at Hastings with its perpetual visitors.

We rode often to the Chiswick Gardens and got specimen flowers to paint; were also often at Kew, and once when there Sir William Hooker gave me a hanging bunch of *Amherstia nobilis*, one of the grandest flowers in existence. It was the first that had bloomed in England, and made me long more and more to see the tropics. We often talked of going, if ever my father had a holiday long enough.

Next season we went another way down to Norfolk by Harrogate, where an Indian uncle was drinking the waters. How that place smelt of sulphur! After a month of mountain air we returned to the flats of Norfolk. After that a stormy Parliamentary session succeeded; my father forming one of a small party of Liberals who called themselves then the St Stephen's Club. We let the house at Hastings for that summer (and the two next also) to Count Poutiatine, the famous Russian Admiral, who ran the blockade of the White Sea so cleverly during the war. When the House was up my sister, my father and myself wandered off abroad, starting by way of Jersey for the Pyrenees and Spain, returning in an English ship from Cadiz to the Thames on the 3rd of January 1860.

After my brother married, my father gave up the old house at Rougham to him, and each summer, when the Parliamentary session was over, we three, with our three old portmanteaux (their collective weight nicely calculated under the 160 lb allowed on Continental railways), used to start forth on some pleasant autumn journey. My father loved the deep romantic valleys round the southern slopes of Mont Blanc and Monte Rosa, and there summer after summer we found ourselves walking over easy passes, with just enough of necessaries to be easily carried on an Alpine porter's back.

In the autumn of 1861 we made a longer journey to Trieste, Pola, Fiume, by the Hungarian Lake of Balaton, where grew such grapes as I have never seen elsewhere in Europe, to Pesth and Debreczin. Here we were lucky enough to see the wild humours of a great Hungarian fair, with horse-races, and a superb gipsy band. Then down the Danube and across the Black Sea to Constantinople, Smyrna, Athens, and home by sea to Marseilles.

The winter of 1863–64 was a merry one. We had a succession of nice people staying with us, whom our young cousins in the Croft used to describe with youthful flippancy as "Old Couples without Encumbrances." In the summer of '64 we went to Pfeffers, crossed the Julier Pass to Samaden and Pontresina, and settled ourselves in that paradise of Alpine climbers, the Old Crown Inn.

Our Egyptian journey, long talked of, came off at last, and it was a misfortune that brought it to pass! At the General Election in July 1865 my father lost his seat, being turned out by George Waldegrave, with a majority of just nine votes: after the first vexation we turned our thoughts to utilising the unlooked-for leisure, and started at once for Switzerland. We then worked our way through Austria and Italy to Trieste where we boarded an Austrian Lloyd boat which coasts the Adriatic by Spalatro, Ragusa and Cattaro.

We put in at three Albanian ports to take in mails, but did not land, and thought ourselves excessively ill used when we arrived at Corfu and found ourselves liable to eleven days of quarantine; I am sorry to say my father lost his temper. He had taken his ticket all the way to Beyroot, and now was not allowed to change ships and go on, just because the Corfu people wished to spite the Turks and make something out of the compulsory board of any strangers they could catch, the whole island since England gave it up being more or less bankrupt. My father sent a letter to the Consul-General, who was in England and could not answer, but his sub, Baron D'E., came out to us after six hours of consideration about it, and found my father in a state of bottled indignation that was truly alarming. He made him a speech over the bulwarks, which was not complimentary to the Greeks and their ways, the captain saying "hear, hear!" The poor Baron became intensely civil, and offered us his own yacht to pass our imprisonment

Doum Palms (*Hyphaene thebaica*) and Date Palms (*Phoenix dactylifera*) on the Nile above Philae, Egypt.

in: our time in it was not pure pleasure at first, as it was about the size of a Hastings fishing-boat, and rolled at anchor considerably. When finally they let us go it was contrived that we should be just too late to start in the next steamer to Beyroot, and we had a whole week to wait for the following one. It is not a great hardship to stop in beautiful Corfu, but we felt we had been trapped, and there was something melancholy in seeing the place going so fast to ruin.

The damp heat of October was depressing, and we never felt really free of quarantine till we steamed away from Corfu in the good ship *Germania*, and soon found we were again in quarantine at Syra, but this time for twenty-eight hours only, in our own comfortable ship, anchored outside the harbour. It was difficult to discover rhyme or reason in those Greek quarantines. Our next stop was Smyrna where we renewed acquaintances and experiences from our previous trip. Our late captain and his two under-officers came on shore on purpose to introduce us to the captain of the ship we were to go on in, which they did as "Monsieur my Lord et mees sa fille," without any trouble about finding a name

for us: but it was meant most kindly, and had the effect of getting us well cared for in the ship which took us out of the beautiful bay of Smyrna next day.

Though we stopped some hours both at Cyprus and Rhodes we could not land at either: one island seemed full of date palms, the other of windmills, and the mouth of Rhodes harbour was quite narrow enough for a Colossus to have rested with one foot on either side. Eliot Warburton's dragoman (then landlord of the Oriental Hotel) took possession of us at Beyroot, and carried us off to his house, and very Oriental it was! From my window I saw many flat house-tops, a woman washing clothes on one, beyond her were fig trees and the blue sea.

We left Beyroot, after visiting Damascus, in a most luxurious Russian steamer. At Port Said the captain lent us his boat and we rowed to the end of the great stone jetty, then in process of making at the mouth of the canal, but they would not let us land without four days' quarantine. We landed the

next morning at Alexandria. It was a nasty, mongrel, mosquito-ish place, and we got out of it as fast as we could.

We had brought the great Michael Hamy with us to arrange for our Nile voyage, but found on arriving at Cairo all his proposals were so exorbitant that my father paid him off, and sent him home again, and we settled ourselves in the Hotel du Nil, in the centre of the town, contented to wait patiently in that most entertaining of capitals till something turned up! Our quarters in the German hotel were most comfortable and quiet. I had a room next the landlady's upstairs, with a window looking into the garden or central court, and my father one on the ground opposite, with a tangle of palms, lantanas, hibiscus, poinsettias, jasmines, and roses between us, and this was in the very heart of old Cairo.

At last we found a dragoman willing to take us up the river at a reasonable rate, and we started on the day after Christmas with two fellow-passengers, – a French gentleman in tight boots and a diamond ring, who traced hieroglyphics very slowly, and kept a journal, and Mr. S., a young architect with the R.A. travelling scholarship, and enormous industry. He took a camera-lucida and traced every squared stone with it from Cairo to Aboo Simbel and back: both were most gentlemanly and agreeable companions, dividing the end of the boat between them, while my father and myself had each one of the side cabins. We were excellently fed by an old cook, who said he had cooked for the English for forty years, but who had no visible eyes. The sailors were a happy simple race, and two of the boys were absolutely beautiful, of a bright shiny copper tint, and liked sitting for their pictures. The captain was as black as a coal and very dignified; he was always curled up on the upper deck on a heap of rags and pillows, and took cups of coffee perpetually with the air of a king, occasionally waking up and giving mild orders; he used most courteously to offer me his pillows each time I went on deck.

The weary periods of the voyage were the many days without wind, when we were pushed and dragged against the stream, or hardly moved at all, but then we could get on shore, where one

was sure to find something fresh to see or draw. We were continually sticking in the mud, and then all the men jumped into the water, wearing their turbans but throwing aside all their other rags, to push and pull us off again.

Sioout was our first regular day on shore, here the men made their bread, and we mounted the side of a hill, and saw our first caves, with gigantic figures sketched on the smoothed surface of the rock to guard their entrance, keeping their eyes at the same time on the rich crops of young corn and tares, lentils, cotton, and sugar-cane, which bordered the great river below. Mr. S. and I worked hard all day. We spent ten shillings in the bazaar on the pretty earthenware pots for which Sioout is famous; then we went on, and soon entered the region of the doum palm. Birds also became more common, we had seen troops of pelicans, ibex, storks, and ducks, and now we had abundance of larks and water-wagtails, and lovely long-tailed green birds almost like parrakeets, but smaller. What people mean by calling the Nile uninteresting I never could understand, we always found abundance to entertain us on shore or afloat. The air there was most delicious, but as the wind continued contrary, and we were told it was "about" six miles only to Thebes, we three English decided to walk, with the distant towers of Karnak as our guide, across the fields. Of course, as usual, it was much farther than we expected, but we reached the forest of pillars by eleven o'clock, and took a couple of hours' rest on comfortable flat stones near the great obelisk of Karnak, where I believe we all went to sleep. The air was fresh, but the sun was the sun of Egypt, and we had gone at a most mad pace for some hours and the grandest ruin in all the world could not have kept my eyes open.

We awoke refreshed, and able to take in the enormous size of that temple, and the rich colouring of its ornamentation, and to enjoy walking through an avenue of sphinxes who were apparently waiting for a surgeon to come to set their broken bones and put the right heads on the right beasts. That avenue led us on to Luxor, which is spoilt by a village taken out of it, sand and all, for at that time it was difficult even to trace the temple amidst all its rubbish.

We went on, drag, drag, drag, again, eight miles only in thirty hours. A large American

A Vision of Eden

dahabieh had reached Assooán before us, and had consequently the right of precedence up the cataract. At first the Sheik swore it was too big and could not go, but the owner, being the American Consul, compelled him to try. On the first attempt it was wrecked on a desert island, where it was left until the next day when it was refloated, wrecked again and finally hauled to the top by a mass of shouting imps and demons.

After this all the howlers vanished also, and perfect calm ensued as we floated gently on to the island of the gods, the very *Heiligen Hallen* of Mozart, and where his glorious song, "O Isis and Osiris," ought to have been sung 2000 years before it entered into the mind of the composer to write it. Philae was even more enchanting than I expected, with its wall of rocky mountains and boulders enthroning it all round in almost a supernatural stillness, – not a soul but ourselves remained at the island. We hurried on the next day, leaving our friends to ride back to Assooán, and hastened to the Second Cataract, passing temple after temple on our way; the sunshine and blue sky were always glorious, though the nights were still often cold. We had been looking a long while through glasses at the distant hills for Aboo Simbel, when all in a moment we discovered the four gigantic figures of Rameses II. calmly looking down on us from just above our heads, bathed in the golden sunset rays, and half smothered by golden sand; we all shouted with delight at finding them. We stayed three days at Aboo Simbel, painting and studying the noble temples and figures, which alone well repaid the whole expense and trouble of a Nile voyage: the four great figures of Rameses II. seemed to me the finest monuments in the world, they are carved out of and on the face of the rock; the material, as shown in the fragments of one Colossus that is destroyed downwards to its knees, being friable as Nile mud, though its grains are of indestructible silex, translucent when seen through a magnifying glass. The figures are sixty feet high, and the hieroglyphics of the entablature are as sharp as on the day they were cut.

We drifted slowly down the river from Aboo Simbel, often sticking in the mud, and with all the men in the water pushing and hauling us off again; stopping days too at all the old Nubian temples, with plenty of time to enjoy and sketch them.

Our old pilot afterwards took Mr. Lear up the river, and, according to him, described me in the following words: "This Bint was unlike most other English Bints, being, firstly, white and lively; secondly, she was gracious in her manner, and of kind disposition; thirdly, she attended continually to her father, whose days went in rejoicing that he had such a Bint; fourthly, she represented all things on paper, she drew all the temples of Nubia, all the sakkiahs, and all the men and women and nearly all the palm trees, she was a valuable and remarkable Bint!"

We were not sorry to get back to our old quarters in the Hotel du Nil at Cairo, and to sleep in beds instead of on hard shelves. All seemed just as we left it. The expected English box had not arrived, and we went to Suez to wait for a week and give it more time; the change did my father a world of good, and we wandered on the sands for hours together, picking up an endless variety of sea shells (but all dead).

We enjoyed our three hours of railway journey back to Cairo exceedingly, the air was so fresh and wholesome, and the scenery so curious. My father got depressed and ill again, waiting for the box that never came, and we were glad to leave Cairo for Alexandria, where the Briggs of the former place told us we should be sure to find it, but the Briggs of Alexandria swore he had sent it months ago to his correspondents in Cairo, so we determined to wait no longer shilly-shallying between them, and we ordered it and our two portmanteaus to be sent to Beyroot, and ourselves embarked for Syria.

We stayed a few hours at Port Said, looking at the big stones, and when we reached Jaffa wondered how we should ever get through the surf and rocks and the rogues at the landing-place. We had almost given up trying in despair when the first boat brought our remedy in Hadji Ali, a dragoman we had seen on the Nile, who was just packing off his party for Constantinople, and had all his tents and horses ready to take us on. Of course he was delighted at a fresh start, and we to find a protector in a man whose name was so well known and spoken of, so he piloted us through the rocks, surf, and streets, and out to the beautiful

Papyrus (*Cyperus papyrus*) growing in the Cyane, Syracuse, Sicily.

24

garden plain beyond, where we found his late party taking their last luncheon in his tents, and they gave him the best of characters and to us some food, after which we went to the Consulate to rest through the heat of the day, while they packed themselves off. It took four or five hours of riding to reach Ramleh. Nine more hours took us on to Jerusalem, which we reached in good daylight, sent our horses round, and walked in by the Jaffa gate and into the sepulchre Church itself – such a fair outside but so rubbishy within, though full of picturesque bits. We then made our way through the city and up to our tents on the Mount of Olives, where the cook being drunk was sent down to the guard-house for the night, and the Hadji turned up his sleeves and cooked our dinner just as well as the cook would have done, and then came and gossiped and smoked with my father as usual.

The city and its position were much what I had expected, but the details even more picturesque; our tents were on a ledge of the mountain, all among the grand old olives, and a hundred yards above the garden of Gethsemane, into which we might almost have dropped a stone; the whole line of the city walls was opposite us, and about on the same level, with the famous Dome of the Rock in front; we went all over that exquisite building, the most elegant in the world.

We made our first extended trip to Jericho, the Dead Sea and Bethlehem visiting many holy places before returning to Jerusalem. In the next few weeks the Hadji guided us through Palestine and Syria to Beyroot and back to our first quarters at the Oriental Hotel, where we found our dear portmanteaus, but not "the" box, that never turned up till we arrived in England: luckily the need for it had ceased months before. We were really unhappy at parting from the Hadji, who had been like a dear friend all the time we were with him. We returned by boat to Trieste and crossed Europe back to England.

We did not stay much in London that next season, 1867, but devoted ourselves to the Hastings garden. After this came two more short journeys to the Italian Tyrol and to Mentone and the South of France. Then in November 1868 George Waldegrave's resignation brought on again the worries and work of a contested election. My father and Mr. Brassey came in with a large majority, but a petition to unseat them for bribery was at once lodged by the opposite party. It was a wretched year at Hastings, though all our kind old friends came down to cheer us with their company – the George Normans, Sabines, Benthams, etc. Then my father resumed his old work in Parliament, but his spirit was broken and his health declining. However, the suspense came to an end at last, and after five days' trial Mr. Justice Blackburn dismissed the petition with costs on the 17th of April 1869.

1869. – On the 4th of August we started for Gastein by way of Frankfort. That journey is so full of painful remembrances that I shall make the note of it as short as possible. After a few days' rest at Salzburg we posted on to Gastein, and got our old rooms at the Hirsch. My father grew so strong in a fortnight that he planned walking over the hills to Heiligenblut – eighteen hours! We went up an Alp 3000 feet above Gastein to try our powers. He came back so well that he went up another hill the next day, leaving me to rest at home, but it was too much; his old disease returned. We hastened down to Salzburg, where the doctor advised us to get home. At Munich he arrived in the greatest state of suffering. The people at the inn were kind, and persuaded me to go for Dr. Ranke. At last I got him safe home. He was so glad to be there, and to see Catherine and his friends again, that they would not believe how ill he was. Even his old friend and doctor, Mr. Ticehurst, did not discover it at first. After a last three days of exhaustion and sleep he ceased to live on the 29th of October. The last words in his mouth were, "Come and give me a kiss, Pop, I am only going to sleep." He never woke again, and left me indeed alone.

For nearly forty years he had been my one friend and companion, and now I had to learn to live without him, and to fill up my life with other interests as I best might. I wished to be alone, I could not bear to talk of him or of anything else. As soon as the household at Hastings was broken up, I went straight to Mentone to devote myself to

North American carnivorous plants painted in England. Behind, on the left, is a Californian Pitcher Plant (*Darlingtonia californica*), with in front a Common Pitcher Plant (*Sarracenia purpurea*); on the right is a Yellow Pitcher Plant (*S. flava*) with in front a Venus' Fly-Trap (*Dionaea muscipula*).

A Vision of Eden

painting from nature, and try to learn from the lovely world which surrounded me there how to make that work henceforth the master of my life. I took our old servant Elizabeth with me; her kind care and the entire rest on the sunny Riviera soon restored the physical strength I had lost in so many months of anxiety and trouble. After about two months we drove on along the edge of the sea to Spezzia, and by rail to Pisa.

We went on board the steamer at Leghorn by 4 p.m., but did not leave till the next morning. On the 28th of February the sun rose without a cloud, slightly tinting with rose the fresh snow on the mountains above Palermo. The absolute necessity of custom-house officers requiring tenpenny-worth of tobacco to smoke, not being a new idea to me, I had no difficulty or delay in getting into a carriage with Elizabeth and the three small trunks, and we drove along the beautiful shore and busy streets of the port, to the famous Hotel Trinacria.

That famous landlord, Signor Ragusa, had a strong objection to travelling ladies, and always pretended he had no rooms for them, except on the fifth floor, so to the fifth we went, and gained all the better view, though I should have been just as well content to have had a handle to my door, as well as a fireplace, but I did not complain. The view from my window was really magnificent, though for one who had passed so many years at Hastings, it seemed very odd to look straight out to sea, and have the sun rise at one's right hand and set on one's left. I saw a long fringe of white surf, and beyond that the deepest blue, green, and purple headlands, one over the other, the higher mountains looking like real Alps in their winter dress, the one permanently white point of Etna made quite insignificant as it peeped over, in the far distance, a hundred miles off. My first weeks there were full of pain and suffering, not lessened by my faithful attendant's tears and repinings, she wished she "had never comed" (didn't I wish so too).

On the 8th of March my doctor wrote me "Please don't go to Girgenti, an Englishwoman travelling like yourself with a maid, has just been shot there." I went down to Ragusa, who got into a terrible rage all over, and said, "That woman, he knew all about her! He would like to write to the *Times*, but didn't wish to have his name mixed up with the thing. Why! she wore little curls all stuck round her face like a Frenchwoman." That last argument was a settler! So I soon believed from what he said that I should be quite safe in going to Girgenti. He also suggested I might ask my banker if he were not right. Good old Mr. Morrison entirely agreed with him, and kindly wrote a letter to his correspondent, Signor Pancarmo, to arrange that I should spend a fortnight among the remains of old Agrigentum.

On the 18th of March we steamed out of the beautiful harbour, and saw it for the last time, catching just one peep more of the town over the Gulf of Mondello. At dinner the captain asked me what I was going to do at Girgenti, and if any one expected me there, for it was an awkward place to get to, particularly in such weather, being four or five miles inland, on the top of a high hill, so I pulled out my precious letter for him to read. He said he would see I got safely on shore, and on to my destination, and when we arrived at the port, he introduced me to a tall man, with a beard, the agent of the Florio Company. This gentleman took me and my encumbrances on shore in his own boat, and would let me pay nothing, but found porters, and saw me through the custom-house, depositing me in his office while he went and found Signor Pancarmo's agent, in whose care he left me. That gentleman was even more kind, and insisted upon going himself up in the fly with us to Girgenti, which stands 1240 feet above the sea.

My good friend from the port arranged everything he could think of for our comfort, and when I attempted to thank him, said, "It was always such a pleasure to help a foreigner, and it had been quite a pleasant excursion for him;" then after a while he walked away down the hill again in the rain. After he was gone came Signor Pancarmo himself, not the old man I expected, but quite young and a most finished fine gentleman. He offered to do everything under the sun for me, understandable and not understandable (for I had not the gift of tongues), but one thing he was positive about, that if I had a proper man to look after me, I was safe to sketch as much as I liked in Girgenti; and he put himself, his carriage, and horses all at my disposition, in true Spanish style, and said he would call the next morning and drive me down himself to see the ruins; and arrange for my work afterwards.

Early Days and Home Life

The first day we followed the edge of the cliff, past masses of old wall with hollows for tombs in it and in the rocks beneath, then through the formless masses of mighty stones which are all that earthquakes have left of the huge Temple of Hercules, with its large segments of pillars and capitals sunk deep in the ground; it was almost as great a puzzle to know how it tumbled down as how it had been erected. Then we went over what had been the Porta Aurea to another heap of stupendous stones called the Temple of Jupiter Olympus. Many of the columns here were only rounded on one side, so that the square side faced the interior of the building. The country was then most enchanting (20th March), every bush had its freshest green, even the slow old figs with their invisible flowers were all at their best. The chief guide of the place met us when we first left the carriage, and Signor Pancarmo engaged him to attend on me. Every fine day, as long as I stayed, he was to be at my door at sunrise and accompany me wherever I liked to go until sunset. The people were all most friendly, Elizabeth felt quite happy and safe, walked down every day and enjoyed her picnic in the fields; it was fun to her, but as usual none to me for I could not do good work in the midst of such a crowd of idle people. I spent many days walking through and working in the innumerable temple ruins of the area before taking leave of my good friends. Signor Agostino came to take us on board the steamer, giving the captain particular directions to take care of us. He told us to go to the Hotel d'Italia in Syracuse, and so we did. We two were put into a large room with a tiled floor, and a window at each end, one looking on the court of the hotel, the other on our neighbour's roofs.

On the first day we went out over the drawbridges and through the different gates. Then we crossed the level and now wasted ground, in the midst of which stood one single rose granite column of the ancient Forum, while some others lay near it. The Roman Amphitheatre was a very perfect oval, with a tank in the middle for crocodiles, but like most Roman work unpicturesque. A German gentleman in spectacles joined us there; my instinct told me, here was a person to fraternise with. I did not repent; he had only one day to spend in Syracuse, and wanted to see all he could, but like myself had been unable to secure the help of the only good guide to Syracuse, Salvatore Politi. I, too, wanted to get a general idea of the old city, so was delighted to have an intelligent companion who knew the ground-plan by heart, and all its histories; we ascended together to the Greek theatre from whence one gets the grand general view of the city.

The Ear of Dionysius came next, a grand old vault, in which we tried every description of echo, including a mere sigh, or the crackling of a piece of paper. I heard both at the further end with the greatest distinctness, though at 200 feet distance from the first sound. It was very hot, but we walked on to the shore in spite of it to engage a boat for the Cyane. After much German bargaining we got one for half the price demanded, and started. We made a false dash to get over the sandbank at the mouth of the Anapus, and stuck fast, so that the men had to get into the water and push us off; but our second attempt was more successful, and we floated in on the top of the wave, between the banks, and soon found ourselves amongst the papyrus thickets. After a while the men declared they could go no farther, but my German knew better, and on we went by dint of pushing and pulling and great struggling, through a mesh of tangled weeds of many varieties, which we pressed down and rowed over, and which started up again after we had passed, and closed the way behind us, every leaf resuming its place at once, as if nothing had happened to disturb its serenity. So we fought our way to the very end and source of the river, to the beautiful Cyane, of wondrous blueness and unknown depth, through which by kneeling in the boat I could see quantities of great fish, and down to the very bottom where the weeds shone out like emeralds. I saw one plant of river-weed springing from a slender stem no thicker than my finger, which spread and spread as it came upwards for thirty or forty feet till the top floated on the surface of the magic pool, covering a square of at least three yards, a perfect inverted pyramid of greenness. The Cyane and its fish seemed both enchanted; the papyrus almost met over our heads as we pushed our way back, and the masses of ranunculus, watercresses, and other weeds, appeared almost thick enough to walk on, but we were told there was not a bit of standing ground within a mile of the Cyane.

A Vision of Eden

We landed, and walked across the fields to the two columns of Jupiter Olympus, on some raised and cultivated ground, near which Nikias fought his famous battle against the Syracusans, which seems now to us who live in the times of such terrible death machines, to have been much ado about nothing; Thucydides stating the Athenian losses to have been fifty men! We entered our boat again and passed over a series of swamps, ditches, and treacherous banks, but only acquired a little mud in the expedition, and at last we reached the port, where I said good-bye to a very good specimen of an Austrian gentleman. I have never seen him since. After the first day, Politi took me under his especial guidance, his old uncle, the Cavaliere of Girgenti, having begged him to do so.

On the last day in Syracuse we drove out at half-past five, the town clocks pointing to half-past four. In the next weeks we visited Acrae, Spaccaforna, the caves at Ipsica, Noto, where we did some shopping, Catania and Taormina, a most curious old town, in one of the finest positions of the world. We had secured the best rooms of the little inn, with a balcony hanging over gardens, and beyond them the precipice and deep blue sea; on one side we could see the great theatre, fitted in between its two rocky crags. The little house was perfection, and the only other staying guest a clever Danish artist; we talked a wonderful language to one another – very different on either side, but which we both considered to be pure German. The abundant mosquitoes soon reminded me that May was not the month for Southern Italy. Elizabeth's constant complaints and weariness tired me out, so I took the steamer from Messina to Genoa on the 18th of May, but, finding that it stopped three days in the harbour of Naples before going on, I left Elizabeth and the trunks at an inn, and went on to Rome to see my old friend, Miss Raincock, sleeping at the hotel and spending the two days in wandering about with her.

After my two days' holiday I returned to Naples, picked up my trunks and Elizabeth, and went on by steamer to Genoa, from whence I sent her home, and made my own way up to Monte Generoso for a month's fresh pure air. I painted the lovely wild flowers there continually, and studied the wonderful rolling clouds and their shadows over the great plain of Lombardy, with Milan cathedral shining in the midst of it.

My sister and her husband, J. A. Symonds, joined me there, and would have taken me on with them into the Dolomite country, but I had been so long alone, that constant conversation tired me; I felt inclined to sit down and cry after a few hours, and felt far more alone than before. I was not strong enough to be good company to anybody yet, so they left me to solitude, and I got happier again. I had no difficulty in finding my way home to the flat again, and had to learn to make it my home, though the one person who made home homelike was gone for ever; to learn to live without seeing his smile or hearing his voice was very hard. Friends were most kind, and tried by their extra gentleness and sympathy to make me feel less alone. I persisted in my work, and gave myself few idle moments for mere useless sorrowings, and so the long winter came to an end.

CHAPTER II

Canada and the United States

1871

I HAD LONG HAD THE DREAM of going to some tropical country to paint its peculiar vegetation on the spot in natural abundant luxuriance; so when my friend Mrs. S. asked me to come and spend the summer with her in the United States, I thought this might easily be made into a first step for carrying out my plan, as average people in England have but a very confused idea of the difference between North and South America. I asked Charles Kingsley and others to give me letters to Brazil and the West Indies, his book *At Last* having added fuel to the burning of my rage for seeing the Tropics.

1871. – On the 12th of July I joined my old friend Mrs. S. at Liverpool; the next day we packed ourselves into a comfortable cabin on board the Cunard steamer *Malta*, and moved away westward. It was rough, and a young French officer thought he was dying, sent for Mrs. S. and asked her to take his last will and testament to his betrothed at Boston. He wished her also to ask the captain to stop and let him out, and he would go on by the next ship. He also wanted her to make the steward bring him some pudding he called "by and by," which the latter was always promising and never brought. My friend promised to do all he wished, and that consoled him, and he didn't die.

The very sight of Mrs. S. did any one good; her head was covered with little curls of pure silver, her complexion was very fair, and she wore a purple knitted cobweb pinned on the back of her head, and diamond earrings. She was full of jokes and continual fits of laughter, her quaint American accent making her talk all the more amusing. She had a very pretty, but perfectly useless little French maid, and an enormous quantity of luggage, which was the plague of the little maid's life, for she could never find anything that was wanted, and used to wring her hands and exclaim "quel horreur!" at everything her mistress required.

After the usual scares of fog and icebergs we arrived safely in Boston harbour, and F.S. was soon on board, coming out with the pilot to meet us, and accompany us in through its many islands, crowded with all kinds of sailing and steaming vessels and many pleasure-yachts. He employed his leisure moments in cramming me with stories about the inhabitants, aborigines, etc. etc., till we were sore with laughing. Yankee stories cannot be written; it is the dry peculiar way a clever American tells them that gives them their charm. We drove out to Newton, about six miles into the country, where Mrs. F. and the baby welcomed us, and next day she drove me into the lanes, where I found many new plants; one of the Sweet Gale tribe, called Comptonia, was very common by the road-side, and had a delicious scent; they called it "Sweet Fern," and indeed its leaf had a brown furry back, and was much like our ceterach; its leaves are sometimes dried for smoking instead of tobacco. Large-leaved oaks, white pines, hemlock spruce and arbor vitae hedges, wych-elms and maples, all showed one was not in England.

We finished our drive by a visit to "Jamaica plain" and its famous confectioner: tied up our pony to a post, then went inside and ate the largest

"ice-creams" in a small and perfectly undecorated room. In the evening we dined with the parents of my hostess at Brookline in the midst of a pretty garden: the whole piazza round the house was covered by one creeper – a fruitless vine, a dense mass of foliage. After dinner we strolled through the near garden and plantations to the top of a little hill to see a flaming sunset amid great thunder-clouds, and returned to the piazza for tea, when other neighbours joined us.

After three days Mrs. S. and I settled ourselves in the house she had taken near West Manchester. It was built on the foundation of a fort or tower on the rocks, against which the sea washed on three sides at high water; the rocks were tinted with pink, red, brown and gray, and above high-water mark with soft gray, green, and yellow lichens, wild grass, and scrub: there was no garden, and we wanted none. On the west side the sea ran up into a little sandy bay, the very ideal of a bathing-place; a few steps would take us down into it from the back door. On the east side were holes under the steep rocks, where we could find water at the very lowest tides, and these were almost as easily reached for bathing.

We used to go for a drive of an evening, and bring home great bunches of scarlet lobelia, which they called the "Cardinal Flower," white orchids and grand ferns, smilax, sweet bay, sumach, and meadow flowers, to dress up our pretty rooms. The railway station was about four minutes' walk from our house – a shed with three chairs and a red flag, which we stuck up on the end of a bamboo placed there for that purpose if we wanted the train to stop. Newspapers and letters were thrown out by the guard as he passed; whoever happened to be going that way picked them up and distributed them, tossing ours in at any door or window that happened to be open: we had also a post-box, No. 115, at West Manchester, in which letters were sometimes found and brought over by friends.

It was an idle enjoyable place, but the heat was too dry and glaring for much work. F. kept us supplied with American papers, coming back-wards and forwards himself for a night at a time,

boiling over with jokes and stories, and making us laugh till we cried. The trains used to mark the hours at West Manchester, and on Sundays, when there were none, I never knew how time went, and wondered that we got our meals as usual. The food on those days was always extra good – huckleberry puddings with cream were quite divine, and corn-cakes and chowder, a most glorious compound of codfish, soup, and crackers, not to be tasted off that coast from the St. Lawrence down to the Hudson. One day we went in a train to Lynn, a town which is entirely inhabited by shoemakers; indeed all the country round is famous for that work, as the ground is so dry that agriculturalists get but a poor living from it alone. Nearly every small farmer has a shoe-shop for spare hours and winter work. From Lynn we drove over a mile or two of sandy causeway, with sea on both sides, to the former Islands of Nohant, now a fashionable watering-place. Longfellow was living in the house of his brother-in-law, Mr. A. The latter had invited us to dinner, and then gone out on a yachting excursion, which every one said "was just like him"; but the grand old poet with his daughters was expecting us under the piazza, and his kind sweet gentleness of manner and pleasant talk quite fascinated me. We spent some most delightful hours listening to him, then missed our train, and had to return in a slow luggage train.

The C.F.s were spending the summer near us at West Manchester, and were very good to me when Mrs. S. was away. He was partner of Ticknor, and editor of the *Atlantic Monthly*, and she a pretty poetess who went into floods of tears at the mere mention of Charles Dickens, whose name resembled that of his own "Mrs. 'Arris" in their mouths, and their room was hung all round with portraits of their hero.

I enjoyed my expeditions with him and his wife. He invited me to meet Mrs. Agassiz at a picnic one day, and called for me in his pony carriage, picked her up at the railway station, and drove us to one of the many beautiful high headlands on the coast; then we walked over the cliffs to find a most curious old cedar-tree, perfectly shaved at the top like an umbrella pine by the sea winds, with its branches matted and twisted in the most fantastic way underneath, and clinging to the very edge of the precipice, its

The old Red Cedar (*Juniperus virginiana*) on the rocks near West Manchester, Massachusetts, U.S.A.

The rocks on the coast near the artist's temporary home at West Manchester, Massachusetts, U.S.A.

roots being tightly wedged into a crack without any apparent earth to nourish it. It was said to be of unknown antiquity, and there was no other specimen of such a cedar in the country; it looked to me like the common sort we call red cedar. We sat and talked a long while under its shade. Mrs. Agassiz and I agreed that the greatest pleasure we knew was to see new and wonderful countries, and the only rival to that pleasure was the one of staying quietly at home. Only ignorant fools think because one likes sugar one cannot like salt; these people are only capable of one idea, and never try experiments.

Mrs. A. was a most agreeable handsome woman; she had begun life as a rich ball-going young lady, then, on her father losing his fortune, she had started a girls' school to support her family, and finally married the clever old Swiss professor, whose children were already settled in the world. She made an excellent stepmother as well as travelling companion, putting his voyages and lectures together in such a manner that the Americans had a riddle, "Why were *Agassiz's Travels* like a mermaiden?" "Because you could

not tell where the woman ended and the fish began!" The Professor was a great pet of the Americans, who were then just fitting up a new exploring ship for him to go on a ten months' voyage to Cape Horn and the Straits of Magellan to hunt for prehistoric fish in comfort. She told me much of the wonders and delights of her famous Amazon expedition, and promised me letters there if I went. After a delightful morning we drove on to the woods behind Mr. F.'s house, and found luncheon spread for us, Mrs. T. and her sister, in white aprons and caps, acting servant-maids and waiting on us. Mr. F. let off a perfect cascade of anecdotes, and then I was taken into the house to do my part of the entertainment and sing for an hour, which I grudged much, as I preferred listening; but I suppose they liked it, as one of the ladies wept bitterly.

Another day I went by street-car from Boston to Cambridge, and met two pretty girls, who spoke to me and told me they were the Miss Longfellows. When we got to the end of the journey, their father came and took me for a walk round the different Colleges, and home to have lunch with him in the house Washington used to live in. It was quite in what we English call the Queen Anne style, with plenty of fine trees round it, and large wainscoted rooms full of pictures and pretty things. The luncheon was worthy of a poet – nothing but cakes and fruit, and cold tea with lumps of ice in it; he was a model poet to listen to and look at, with his snow-white hair, eager eyes, and soft gentle manner and voice, full of pleasant unpractical talk, quite too good for everyday use. He showed me all his treasures, and asked me to come and stay with them if I returned to Boston, after which he showed me the way to Mrs. A.'s house. I found her and the Professor even more to my mind; he spoke funny broken English, and looked entirely content with himself and every-body else. They showed me photographs and told me of all the wonders of Brazil, and what I was to do there, then gave me a less poetical dinner. Then Mrs. Agassiz took me to the Museum

and made Count Pourtalez take us up to the attic to see the most perfect collection of palms in the world (all mummies), intensely interesting, as illustrating the world's history. Mrs. Agassiz showed me the great sheath of one of the flowers, which native mothers use as a cradle and also as a baby's bath, it being quite water-tight. The flowers of some of the palms were two to three yards long. She said, though she had wandered whole days in the forests, she had never seen a snake nor a savage beast. One day she heard a great crashing through the tangle and felt rather frightened, when a harmless milk-cow came out. After seeing the palms she caught a German professor and made him show us a most splendid collection of gorgeous butterflies: I never saw any so beautiful; they were all locked up in dark drawers, as the light faded them. Then came corals and madrepores.

Before we left the coast the sumach was turning

Autumn tints in the White Mountains, New Hampshire, U.S.A.

geranium colour, and one little hill near looked as if it were burning with it. The red berberry bushes were also a beautiful deep tint. I found lots of creeping moss with corkscrew shoots above ground, and the root creeping underneath for twenty yards or more, sending up its pretty branches at every joint.

At last Mrs. S. made up her mind to the long-talked-of journey to Canada, and we started with an enormous quantity of luggage. We travelled through Alton, Lake Winnepiseogee, Quebec, the Falls of Montmorency and Montreal. At the end of the journey we rushed through Kingston and Toronto, and arrived at ten o'clock at night to find comfortable rooms at the Clifton Station Hotel, kept by an old Swiss courier, "Rosli."

I was fairly tired out the next morning, but the quiet homely quarters suited me, and I determined to stay quiet at least a fortnight so as to enjoy and sketch Niagara at my leisure. It was so cold that I was glad of the two miles' walk back from the falls, after getting half-frozen over my sketching all day. The season was over, the big hotels nearly closed, and the wooden shops were being moved away bodily on rollers to other and warmer quarters for the winter. But the natives took the greatest interest in my work, and made several offers to buy it. A woman at a toll-gate, near which I had been sketching a marvellous group of coloured maples two mornings running, refused to let me pay toll when returning in a carriage, as she said I worked too hard for her to take anything from me.

The falls far outstretched my grandest ideas. They are enormous, the banks above and below wildly and richly wooded, with a great variety of fine trees, tangles of vine and Virginian creeper over them, dead stumps, skeleton trees, and worn rocks white with lichens; the whole setting is grand, and the bridges are so cobwebby that they seem by contrast to make the falls more massive. From my home I could walk along the edge of the cliff over the boiling green waters all the way to the falls, and if they had not been there at all I would willingly have stayed to paint the old trees and water alone. Mr. Rosli gave me wonderful accounts of the falls in winter, when great masses of ice came down from Lake Erie, got jammed between the rocks and banks, and gradually froze the water between them, then more ice slipped

under and it was lifted up like a bridge; he said it was a most marvellous sight, and he had known carriages driven across on the ice under the bridge, but that did not often happen. It is much milder at Niagara than in Lower Canada, grapes and peaches ripen better; the old arbor-vitae trees are splendid, as scraggy as any old silver firs, and the oak trees are drawn up into grand timber, the trunks rising without a branch for over fifty feet. It was difficult to choose out of so many subjects where to begin. The Horseshoe Fall tempted me much, standing close to its head, with the rapids like a sea behind, and the rainbow dipping into its deep emerald hollow; the tints were endless in their gradations, and delicious, but I got wet through in the mist.

Another tempting bit was below my home, looking down on the whirlpool, where the savage green boiling water seemed piled up in the centre like some glacier; there were foregrounds of great arbor-vitae trees almost hanging in the air like orchids, with long twisted bare roots exposed against the edge of the cliff, from which all the earth had been washed. The rapids about Goat Island on the American side were also full of wonders. One day it blew such a gale that I had to sit down and hold on tightly to the bars of the bridge on returning; no carriages attempted it that day. There are thirty-five minutes difference in the time on the two sides of that bridge, and passengers are charged 40 cents for walking over.

The Head Guide of the Falls, who came out from Scotland forty-seven years before, patronised me, and told me if I got chilled at any time just to go and ask his missus to give me a good cup of coffee, it 'ud do her heart good to make it for me. He showed me some lovely views at the bottom of a rickety old tower about seventy feet high, with a corkscrew staircase winding round one noble pole in the centre. The tower is fastened half-way down to the side of the cliff by an iron bar; it shakes and trembles with every step of persons going up or down. When I had settled to my work on the boulders below, between the two huge roaring falls, I began to think what would happen if it were to tumble down, and they were to forget my being there. But I had plenty of company passing and repassing after the first morning hours. Strange figures in suits of yellow oilskin came and looked at me at intervals. When I

View of both falls at Niagara, North America.

had got my sketch in, and myself sufficiently soppy, I went farther under the spray of the American fall and saw three quarters of a circle of rainbows on it, and watched the yellow oilskin people scrambling over the huge boulders in and out of the clouds of spray; they had left the paths and bridges, and were tempting death from the mere love of danger, but with that steady nerve and strength which showed them to be beef-fed islanders, and fit compatriots of Tyndall.

While at Clifton I got a letter from two old Norfolk servants of my father, John and Betsy Loades, who had settled in Pontiac in Illinois. He was one of our Rougham boys, and eventually became our coachman, and could turn his hand to anything; he married our cook, and they had both helped me to nurse my father, and would have no other master when he died, but emigrated with their two girls to America. Betsy Loades now wrote, "that after knocking about at Chicago and other places, they had settled at Pontiac on the Vermilion River, and John had work for the winter on a new railway; they liked the place, and hoped to buy land there in time; that a man might rent a good farm (as soon as he was able to get a team and a few machines to work his land with), on which he could make a great deal of money in a few years to provide for old age."

I could not resist the temptation to go and see

them and something of the prairie country besides, so I left my portmanteau with Mr. and Mrs. Rosli, and went off with my small hand-bag and sketch-book to Toronto, to refill my purse and see my cousin Dudley's friends, Judge G. and his family.

The G.s put me into a Pullman car the next morning, and for 75 cents extra I was in solitary glory till 8 o'clock at night, with only the occasional society of three guards and the black man in charge, who now and then came up to say "Wall, how are you? Quite comfor-table?" At Sarnia we had to cross a ferry to the other side of the St. Clair river, and then get into crowded cars the very reverse of the solitary luxury I had had all day.

We were turned out at Detroit, and the guard warned me to be quick, for the Great Western Express would barely stop, and indeed it was rolling on again before I was fairly inside the door. At last we came to the white sand-hills round the great Lake Michigan. It was pouring with rain, and we went over endless bogs. Damp farmers and their families came in and out all the way. We reached Joliet just in time to see the train I wanted to go on by leaving the station and going slowly out of sight. There was no other passenger train till eight at night, so I decided to go by the freight train in two hours, and having had nothing but

A Vision of Eden

coffee and biscuits for twenty-four hours, I went to a baker's and had more coffee and bread for dinner. The nice baker's wife was rather doubtful about the freight train. If it was full they might be a rough lot, she said; women hardly ever went that way. However, I risked it, and was most comfortable, and hospitably treated in the one carriage, the guard's van, with three windows and three doors, a great stove in the middle, a divan all round, and an arm-chair and two stools, quite a cosy little room. At last we reached Pontiac about ten o'clock, and the guard lighted me on to a pavement (boards), and told me to follow "that gentleman," he would show me the way to the hotel; so I tramped after him through the mud and rain, and he (a mere labourer) showed me the way most kindly, found the landlord out of a crowd of people in the shop below, and I soon had a capital supper and bed, which, though only a bag of straw, was a great luxury.

The next morning I started to find "Big John." First I went to the post-office. The postmaster had never heard his name; he was a new-comer himself; I had better ask next door. Next door was a shoemaker and watchmaker combined, and he had his eye fixed in a magnifying glass over the anatomy of a damp clock. He was a thorough Englishman, and remembered both John and his watch, and described them, but could not say where he lived. Other gossips dropped in who also knew him, but not where he lived, and they advised my going to the new railway where he worked. So I tramped on again through the rain and the mud outside the town to the new station, and the stationmaster told me if I walked up the line I "should find him in a fur cap," which I did, and John straightway took off that fur cap and dashed it on the ground, and said, "Laws, if that beant Miss Maryhand!" Then went and told his "boss" he must have a holiday, and took me home to see Betsy. Poor fellow, he was the ghost of his old self. So thin from constant attacks of dumb-ague, but he said he meant and hoped to live it down, and thought he should get on in time. The ague only attacked new-comers. He had a good boss, and got nearly two dollars a day. His wife made the best of everything in the same way. They took me for a walk through the fields; all the land was black but rich, and magnificent crops of corn and even grapes grew without manure. We saw grand fields of Indian corn-stalks on which the cattle feed in winter, and weeds as high as the corn – iron-weed and cockle-berries. The farming was most wasteful; one field was quite white with the shed beans left on the ground.

On the day I left the landlord's son brought his buggy to drive me to Chenoa to catch the twelve o'clock train to Logansport. Another long day was passed in skirting the southern shore of Lake Erie. I passed more pretty country, full of snake-fences and ague, and saw the big lake with waves like a real salt sea. Toledo seemed a busy manufacturing place, and Cleveland was even bigger. They are monstrous places, as black as Manchester. At Buffalo I had to drive in an omnibus to another station, and then on again by rail. I was put out on the American side of the suspension bridge, and had to walk across in the dark starlight over the roaring river, while a train rolled over my head at the same time, shaking every iron bar. The Canadian toll-taker rubbed his eyes, and said "I was wondering what had become of you!" and refused with indignation to look in my bag for tobacco. Mr. and Mrs. Rosli shook me by both hands, and sent me up buttered toast for tea; my little room looked quite like home again, and, but for the cold icicles hanging round the window, would have tempted me to stay on. I took a last stroll to say good-bye to all my pet views of the mighty waters.

I started by the night train so as to get to Albany by daylight and see the Hudson river afterwards. The Hudson seemed to me like a very mild Rhine minus the castles. A clever talking woman travelled part of the way with me. She was very good-natured, and on our arrival at New York put me into a fly with my luggage. That quarter of an hour's drive cost me eight shillings! New York is not cheap. At the Hofmann House they gave me a very good room, looking on a deep well, with windows all round it, hot and cold water laid on, cupboards and all sorts of nice furniture, and five dollars printed on the door, a pound a day for room alone! My food came in as I ordered it, from a restaurant, and was good and cheap.

I found a heap of letters to answer, and took a day's rest; then I went to call on Doctor Emily

View of the American fall from Pearl Island, Niagara, North America.

38

A Vision of Eden

Blackwell, waiting in her back room till all the patients had had their turn, when she came out and we had a long talk.

I went home to luncheon and packed my bag, returned to the ferry, and by it to the railway. In another half-hour Mr. S. met me: he looked an ideal of benevolence and philanthropy, one of New York's most respected merchants; his carriage and ungroomed horses were shabby to the last degree; he drove himself, and I held the old patched reins while he did various errands in A., which was one of the oldest settlements in America, and quite an historical place; it was situated on an arm of the sea which looked like a lake, near which was the house of Eagleswood, a comfortable but not showy dwelling. That morning the old gentleman with his son and a fisherman had saved the lives of three boys from drowning, their boat having turned over in a squall.

Instead of going straight back to New York I got out at Newark, and went by horse-car to Orange, where I left my bag at the office, and walked off in search of Sydney Clack, the young gardener my father had had at Hastings, who also emigrated after his death. After a few false starts, I found my way through the woods to Mr. M.'s house, prettily placed on a long wooded hill with a view of New York and the Hudson river fifteen miles off. I found Sydney in his green-house, looking well and happy, with two or three men under him, and forty dollars a month, his board and lodging; he had several nice greenhouses, and beautiful flowers. He asked me if I would go in to see Mrs. M., who had told him to be sure to ask me to go in if I came. She was very hospitable, and pressed me to stay and come again.

When I got back to the Hofman House, I found a kind invitation from Mr. Church (the first of living landscape painters) to come and see him and his wife at their cottage at Hudson. They never got my answer, and I missed my train, and only reached Hudson in darkness and rain at half-past eight. I got into a fly and told the man to drive to Mr. Church's. "All right," said the man, and put two other persons in (a way they have in the States), and on we went. Presently he opened the door: "Where did Mr. Church live?" How should I know? but the other passengers said six miles off; so I went to an inn for the night, and then started in

a buggy, and met Mrs. C. in the road coming to hunt for me, and she took me home. She and her husband were quite ideal people, so handsome and noble in their ways and manners.

In my own tiny bedroom were three pictures in oils – one of the Horse-Shoe Falls of Niagara, a study of magnolia flowers, and one of some tropical tree covered with parasites. They had imported two white asses from Damascus for Mrs. C. and the children to ride, and had also a gray South American donkey, quite curiosities in the United States, where the animal is almost unknown.

On my return I found a note from Mrs. M. (Sydney's mistress) asking me to come and stay with them, and as I liked to see as much as I could of life in America, I went back to Orange in the pouring rain, and Mr. M. met me at the station. We had a late dinner in the English style, with wine (which one does not often see on the table).

At a quarter past seven we had breakfast, and the boys and their father went off for their day's work by rail, and after a gossip over the plants in the greenhouse with Sydney, Mrs. M. drove me through the park, not long before a natural forest, then sold in lots to rich merchants for building and farming. A speculator had made seven miles of winding road through it. The views were fine; there were wild rocks, glens, ferns, wild azaleas, and dogwood which in spring is covered with white flowers (like snow, they said). Some of the houses were pretty; some odd and unpractical ones had been designed by the original speculator when under the influence of "the spirits," who did not seem to excel in architecture, I thought.

On my way back I found the ferry crowded with smartly dressed people thronging to welcome the young Grand Duke Alexis of Russia, and

Wild flowers from the neighbourhood of New York. In the foreground is Jack-in-the-Pulpit (*Arisaema triphyllum*) together with the Large Yellow Lady's Slipper Orchid (*Cypripedium calceolus* var *pubescens*), Small Solomon's Seal (*Polygonatum biflorum*), False Spikenard (*Smilacina racemosa*) and Wild Geranium (*Geranium maculatum*). In front, on either side, are flowering branchlets of the Black Huckleberry (*Gaylussacia baccata*).

A Vision of Eden

I began to wonder how I should get on, when a lady told me to follow her, which I did, and she showed me the best omnibus.

Mrs. Botta was my best friend in New York, and soon made me leave the hotel and take up my quarters in her house. She was a most charming and cultivated person, had written one or two books on education, and brought up more than one set of orphan children just for love, having none of her own. She lived with her mother, was a lady of independent fortune, and had married an Italian professor.

Mr. De Forest came one morning and took me to the Johnstone Gallery, a most exquisite collection of pictures. The Great Niagara and a beautiful sunset scene in a swamp by Church were there. The latter is a wonderful picture. The four ages of life by Coles, and splendid Mullers and Cromes were there too. Every picture was a gem.

I went on to Washington with a parcel of clam-sandwiches in my bag, which Mrs. B. made with her own dear hands, cutting a roll in half, buttering it, and putting the odd fishy things between; they come out of bivalve shells something like our scallops.

I did not stop at Philadelphia, but went straight to Mrs. Russell Gurney at Washington at 1512 H. Street. Mr. G. had given up his comfort at home to come and try to settle the *Alabama* question, and was very weary of the task. Month after month passed, and still nothing was settled. It was very cold, and I felt sorry for my friends there; they were most kind and hospitable to me. The first day I went to see Dr. and Mrs. Henry in their pretty museum building, built of pink stone with much-ornamented round archways, and creeping plants over it, and Miss H. showed me many interesting things. There was a large collection of birds' nests, and one trunk of a tree with holes made all over it by a Californian woodpecker in order to pick out its own pet grubs; then the chipmunk or squirrel puts the acorns in, which another bird steals again. We saw also the last of the auks, with its one odd egg; and a horrid little baby mummy which was tossed out of the middle of the earth by an earthquake in South America, and was supposed to be one of the very oldest of dead human beings.

We had a party at home of diplomatic people who discussed some of the new American ways. The young ladies have clubs among themselves, and give parties on alternate nights during the winter, every "Miss" bringing a gentleman. Mamma only has the privilege of giving the supper, appearing while it is being eaten, and retiring afterwards. Papa is allowed the privilege of paying for it, and does not appear at all. These girls go out to other people's houses under the escort of some young gentleman. Pas and mas have a dull life of it in U.S. society. When a man calls at a house he never asks for the mother, only for the girls, and the mother does not appear; if she did she would be snubbed, and made to know her place very quickly.

I had a card brought me the next morning, "the Secretary of State" and Mr. Fish followed it, to whom I had a letter of introduction. He was a great massive man, with a hard sensible head. He said he would call for me in the evening, and take me to the White House. So at eight o'clock in he came again after another big card, I being all ready for him in bonnet and shawl, and in no small trepidation at having to talk tête-à-tête with the Prime Minister in a small brougham. However, I found there was no need as he did it all himself. We were shown in first to the awful crimson satin room which Mrs. G. had described to me, with a huge picture of the Grant family all standing side by side for their portraits. Then we were told to come upstairs, and passed from state-rooms to ordinary everyday life up a back staircase, which was the only means of reaching the upper storey allowed by the architect of seventy years ago. We were shown into a comfortable library and living-room, where a very old man sat reading the newspaper, Mrs. Grant's papa, who did not understand or hear any of the remarks Mr. Fish or I made to him. Then came Mrs. Grant, a motherly, kind body; then at last came the President, also a most homely kind of man.

We at first sat rather wide apart, and I had more of the talk to do than I enjoyed, and felt like a criminal being examined till Mrs. Grant hunted up a German book full of dried grasses to show me, and the poor withered sticks and straws brought dear Nature back again. I put on my spectacles and knelt down at Mrs. Grant's knee to look at them. They began to find out I was not a fine-lady worshipper of Worth, and we all got chatty and happy. Mrs. Grant confessed she had no idea "Governor Fish had brought me with him, or she would not have let me upstairs, but didn't mind now''; and she told me all about her children; and if I had stayed long enough would, I have no doubt, have confided to me her difficulties about servants also. The two big men talked softly in a corner as if I were not there, and I watched till Mr. Fish looked like going away, and then I rose. They were all so sorry I could not stay the winter there, and hoped I would come again, etc. etc., like ordinary mortals; and Mr. Fish showed me a water-colour drawing of the Grants' country house, took me into a blue satin room, which he said was very handsome, and conducted me home again.

I wondered if Gladstone or Dizzy would have taken as much trouble for the daughter of an American M.P. who brought a letter from the Secretary of an English embassy.

The next morning I found a big envelope with a huge G. on it, and a card inside from the President and Mrs. Grant asking me to dinner that night. The Gurneys had another, so we went in state and were shown into the blue satin oval room, well adapted for that sort of ceremony, and the aide-de-camp General Porter came and made himself most agreeable to us. Then came two Senators and the Secretary of Foreign Affairs, and then the President and his wife arm in arm, with Miss Nelly and a small brother, and grandpapa toddling in after. He had an armchair given to him, and General Grant told me he was so heavy that he had broken half the chairs in the house, and they were very careful about giving him extra strong ones now. After a terrible five minutes, dinner was announced, and to my horror the President offered me his arm and walked me in first (greatness thrust upon me). I looked penitently across at Mrs. Gurney, who looked highly amused at my confusion, and did not pity me in the least. I was relieved by finding the great man did not care to talk while he ate, and General Porter was easy to get on with on my other side. He seemed to know every place, inhabited and uninhabited, in America.

He gave me some curious accounts of the few remaining Indians, some of whom are as near animals as mortals can be, too lazy to look for food till the strong pangs of hunger seize them, when they sit in a circle and beat down the grasshoppers with whips, gather them up and crush them in

their hands, eating them just as they are, and then sleep again till the next fit of hunger seizes them. The President drank tea with his dinner, and had every dish handed to him first. He seemed an honest blunt soldier, with much talent for silence. His wife had a funny way, when shaking hands with people, of looking over their heads, and appearing to read off their names out loud from some invisible label there. I was taken out from dinner in the same distinguished manner, being made to stop in the red satin room and admire the

Wild flowers from the neighbourhood of New York. In the front, on the left, is the Naked Broomrape (*Aphyllon uniflora*), with behind the scarlet and yellow Wild Columbine (*Aquilegia canadensis*) and in the centre the pink Lady's Slipper Orchid (*Cypripedium acaule*). On the right is a Pinxter Flower (*Rhododendron nudiflorum*) and Stagger Bush (*Lyonia mariana*).

family portraits and the youngest boy in a Grant tartan and kilt. I asked the President if he did not mean to go some day and hunt for his relations in Scotland, but he said he had quite lost all trace of them, four generations of his family having lived in America, and that he was "raised" in Ohio; and he sat down by me and was quite conversational. I told him about my visit to Pontiac. He said it was quite possible to live down ague, and that after seven or ten years of cultivation the prairies ceased to be unhealthy. How sad it is that the first brave men who make the country must be the victims to its climate. The G.s were quite surprised (as I was) at the fuss the Grants had made about me, as they never gave dinners (they themselves had only dined there once before, when the High Commissioners first went over). I could not think what I had done to deserve all this; but after I left it came out. Mrs. Grant talked of me as the daughter of Lord North, the ex-Prime Minister of England. I always knew I was old, but was not prepared for that amount of antiquity.

We drove out to Arlington, the late home of General Lee, a tasteless building of would-be classical style in a beautiful situation, with distant views of Washington; a one-armed ague-stricken soldier was its only inhabitant.

It blew a perfect gale of wind whilst we were in that dismal place, and we were glad to get out of it again. The large-leaved oaks were still holding on tightly to their brown-papery leaves, and kept up a continual crackle and rustle. In and about all the great towns of the States I saw little houses built for the accommodation of sparrows; the birds had been imported from England to get rid of a caterpillar which had been infesting the trees and eating up everything. The sparrows seemed to take kindly to their new homes and diet but it was still a problem how they would endure the winter. The Potomac was frozen, and people were skating everywhere.

Miss H. brought me some beautiful dried specimens of the creeping fern with leaves like ivy which only grows at some place in Connecticut; it had been so much picked that a law was made and a heavy fine imposed on any one taking more of it.

One morning we went to the opening of Congress; we drove to the Capitol after breakfast, a really handsome white marble palace with a large dome over its centre; then wandered on up

Study of Gulf Weed (*Sargassum vulgare*), a seaweed that thrives in the warm Gulf Stream of the Atlantic.

and down, asking our way till we got to the gallery reserved for diplomats in the Lower House, and were told to take the front seats by Mr. P. the publisher to whom I had a letter, and who seemed to be a universal busybody and most important personage. The House looked twice as large as our Houses of Commons; all the names were read over to "the Bar of the House" (though there was no Bar). The oaths were decidedly calculated to keep truth-telling Southerners out, as they swore they had never counselled nor helped in any rebellion against the government of the U.S., etc. etc. There were two black M.P.'s particularly well dressed (not a general fault in the assembly), and there was a very ample supply of bald heads, as well as some preposterously young-looking men. There was a female reporter among the others in gold bracelets and a tremendous hat and feathers; the messengers were all boys, who dashed about continually amongst the members below, sitting between whiles on the steps of the Tribune. After a while a quorum of both Houses was declared, and a message sent to know if the President had anything to say to them. The House adjourned for half an hour, so we went out, and afterwards into the Upper House, where we stayed to hear the President's Message read, as it was done at the same time in both Houses. The Senate House looked dull after the other, and the Message was very long, it took nearly half an hour to read. The boy-messengers there were smaller than in the other House, some of them did not look more than eight years old; they sat on the steps of the Speaker's platform, and were very ornamental, reminding me of little boys in the foreground of old Italian paintings. We saw Sumner there, with a grand head; Butler, too, I saw in the other House. The lower one was filled with desks standing in pairs, and as they were distributed by lot, people who did not love one another must occasionally have been rather closer than they liked. There were only seventy altogether in the Senate, and each senator had his own desk, armchair, and spittoon. Both Houses seemed to have newspapers and periodicals on their desks, and could read through dull speeches openly, without having to creep up to dark corners of the gallery, as I have seen some highly respected M.P.'s in our own House (with Dickens's last number in their hands). Tobacco

and cigars were selling in the lobbies. The central hall and passages were lofty, and full of fine marbles and frescoes.

On my return to New York, Mrs. Botta took me to some private theatricals in a friend's house. She had a real dramatic gift, and could make her audience either cry or laugh as she pleased. Roasted oysters are a great supper-dish at American evening parties, one oyster being as much as any one could eat at a meal. I think no food is better than those huge American shell-fish.

It was very cold before I left New York, and some snow had fallen, which made me very happy to go on board the Jamaica steamer on the 15th of December. I had a fearful cold in my head, and was nearly frozen to death, but we got warmer every day, and I always better as it grew warmer. We were soon amongst the mysterious festoons of floating gulf-weed. Even the sea-water was warm, and it looked such a solid black blue, and the weed as gold or amber on it, with the long streaks of floating white foam over all.

CHAPTER III

Jamaica

1871–72

IN THE WEST INDIES at last! Christmas Eve!

We passed Watling's Island and Rum Key, and after steaming through the crooked island passage we had a most exquisite sunset, the gold melting into pure blue so suddenly, and yet so softly, that one could hardly say where the beginning or ending of either colour was. The next day we were within sight of Cuba, and the sunset had all the soft colours of a wood-pigeon's breast. The approach to Port Royal, with its long spit of sand and mangrove swamp, and then into the calm bay of Kingston beyond, was intensely exciting. Every tree was of a new form to me, the grand mountains rising gradually up to 7000 or 8000 feet beyond, all creased and crumpled with ins and outs, like brown paper which has been much used.

I landed entirely alone and friendless, but at once fell into kind helpful hands. A young Cuban engineer appeared from the moon or elsewhere, hunted up my luggage, paid my carriage and porters (for I had only American money), and saw me safe to the inn. The next morning the landlady took me at daylight to see the opening of the new market. It was Christmas Day, and all the negresses went in their gayest ball-dresses. On our return I found Dr. C., who insisted on carrying me off to stay. One day Mrs. C. took me a drive up the Newcastle road; when it came to an end we walked on, and I saw a house half hidden amongst the glorious foliage of the long-deserted botanical gardens of the first settlers, and on inquiry found I

A road near Bath, Jamaica, lined with palms (probably *Euterpe* sp.), Bread-fruit (*Artocarpus altilis*) and Cocoa (*Theobroma cacao*).

could hire it entirely for four pounds a month. So I did hire it, and also furniture for one bedroom. Mrs. C. found me an old black woman, Betsy, to look after and "do" for me. There was also a man attached to the house, old Stewart, a coal-black mortal with a gray head and tattered old soldier's coat, who put his hand up to his forehead with a military salute whenever I looked at him. I gave these old people six shillings a week to take care of me, and felt as safe there as I do at home, though there was not a white person living within a mile.

From my verandah or sitting-room I could see up and down the steep valley covered with trees and woods; higher up were meadows, and Newcastle 4000 feet above me, my own height being under a thousand above the sea. The richest foliage closed quite up to the little terrace on which the house stood; bananas, rose-apples (with their white tassel flowers and pretty pink young shoots and leaves), the gigantic bread-fruit, trumpet-trees (with great white-lined leaves), star-apples (with brown and gold plush lining to their shiny leaves), the mahogany-trees (with their pretty terminal cones), mangoes, custard apples, and endless others, besides a few dates and cocoa-nuts. A tangle of all sorts of gay things underneath, golden-flowered allamandas, big-nonias, and ipomoeas over everything, helio-tropes, lemon-verbenas, and geraniums from the long-neglected garden running wild like weeds: over all a giant cotton-tree quite 200 feet high was within sight, standing up like a ghost in its winter nakedness against the forest of evergreen trees, only coloured by the quantities of orchids, wild

47

pines, and other parasites which had lodged themselves in its soft bark and branches. There was a small valley at the back of the house which was a marvel of loveliness, bananas, daturas, and great *Caladium esculentum* bordering the stream, with the *Ipomoea bona nox*, passion-flower, and *Tacsonia thunbergii* over all the trees, giant fern-fronds as high as myself, and quantities of smaller ferns with young pink and copper-coloured leaves, as well as the gold and silver varieties. I painted all day, going out at daylight and not returning until noon, after which I worked at flowers in the house, as we had heavy rain most afternoons at that season; before sunset it cleared again, and I used to walk up the hill and explore some new path, returning home in the dark.

After about a month of perfect quiet and incessant painting at the garden-house, people began to find me out, and the K.s rode down and made me promise to come to their cottage for a night. Their home was a thousand feet higher than mine, with a most lovely view, and tufts of bamboo all round it, the first large specimens I ever saw; they made me feel in another world among their rattling, creaking, croaking, cork-drawing noises. Some of the canes must have been fifty feet high, thicker than my arm, and full of varied colour. There was a pretty garden, crammed with strange new plants. The cysak, which they told me was the sago palm, was very thriving. I began a sketch of the bamboos the next morning, and then went on a mile along the ridge to stay with Captain and Mrs. H. and the old deaf General Commander-in-Chief, in a bare tumble-down old house, supported by two weird old cotton-trees and a sandbox-tree, built on the very edge of the precipitous wall of the valley.

Captain Lanyon came up with the Governor's orders that I was not to go down the hill without coming to stay at Craigton; but I wanted more clothes and paints, so Captain H. promised me a horse at six the next morning to take me and bring me back; but when I got up I found the house like a tomb, not a creature stirring.

I got out of my window, only a yard above the ground, and went down to the stable: all asleep too, and the sun rising so gloriously! I could not waste time, so took my painting things and walked off to finish my sketch at the K.s. They sent me out some tea, and I afterwards walked on down the hill, among the ebony-trees and aloes, home. I passed one great mass of the granadilla passion-flower, with its lilac blossom and huge fruit, which is most delicious, and almost more than one person can eat at a time. I found a Kingston doctor and his family had accepted my offer of rooms for a change, and had come up, furniture and all, for a week to a corner of my vast domain. So after a rummage and a bath I went up the hill again, and old Stewart carried my portmanteau on the top of his head as far as the little collection of cottages at the foot of the Craigton mound.

There was one of the great cotton-trees close to the path, and I went on zigzag, returning continually to the huge skeleton tree, and thought I should never get above it. I reached Craigton just after sunset. The house was a mere cottage, but so home-like in its lovely garden, blazing with red dracaenas, *Bignonia venusta*, and poinsettias looking redder in the sunset rays, that I felt at home at once. The Governor, Sir John Peter Grant, was a great Scotchman, with a most genial simple manner, a hearty laugh, and enjoyment of a joke. I begged to be left off formal breakfasts, went out after my cup of tea at sunrise as I did at home, and worked till noon. My first study was of a slender tree-fern with leaves like lace-work, rising out of a bank of creeping bracken which carpeted the ground and ran up all the banks and trees, with a marvellous apple-green hue. In the afternoon I could paint in the garden, and had the benefit of the tea and gossip which went on near me, sitting under a huge mango, the parson, his wife, and people coming up on business from the plains with three or four neighbours and idle officers from Newcastle.

One day the captain started Agnes Wilberforce and myself on two horses with a groom for Newcastle, where he had arranged that Dr. S. should meet us and show us the famous Fern Walk. It was a glorious day. We rode up the steep hills straight into the clouds, and found rain in the great village of barracks, but we went on in spite of it. At last we turned into the forest at the top of

Foliage and fruit of the Akee (*Blighia sapida*) which is native to West Africa, but widely cultivated in the tropics for its edible arils, as here in Jamaica.

48

A Vision of Eden

Above: View from the artist's house on the Newcastle Road, Jamaica.

the hill, and rode through the Fern Walk; it almost took away my breath with its lovely fairy-like beauty; the very mist which always seemed to hang among the trees and plants there made it the more lovely and mysterious. There were quantities of tree-ferns, and every other sort of fern, all growing piled on one another; trees with branches and stems quite covered with them, and with wild bromeliads and orchids, many of the bromeliads with rosy centres and flowers coming out of them. A close waxy pink ivy was running up everything as well as the creeping fern, and many lycopodiums, mosses, and lichens. It was like a scene in a pantomime, too good to be real, the tree-fern fronds crossing and recrossing each other like network. One saw dozens at one view, their slender stems draped and hidden by other ferns and creeping things. There were tall trees above, which seemed to have long fern-like leaves also hanging from them, when really it was only a large creeping fern which had found its way over them up to the very tops.

Gertrude S., the Attorney-General's sister, soon rode down to see me; she lived only half a mile from Craigton, and was the person I liked best in Jamaica. As a young girl she had been taken out with her brothers and mother by a stepfather to Australia, where she had had no so-called "education," but had ridden wild horses and driven in the cattle with her brothers; had helped her mother to cook, wash, make the clothes, and salt down the meat; and till seventeen she could barely read; then her mother's health broke down, and she accompanied her back to England, nursed her through a long illness and educated herself. Her eldest brother soon took great honours at the Bar, and was sent for by the Governor to help him in starting the new Constitution of Jamaica. I never knew a more charming brother and sister! so entirely happy together, and helpful to one another. Gertrude had taught herself German, French, and Italian in those few years, and still read much, though she did all the finer kinds of cooking with her own hands, and saw her horse and cow fed regularly.

Some of the wild fruits were very good, though the English seldom eat them. The "soursop," or custard-apple, was an especial favourite of mine; it was a green horny heart-shaped thing growing close to the stem of the tree, with a creamy pulp

Opposite: Foliage and flowers of *Alpinia zerumbet*, from Eastern Asia, and a pair of Streamertail Humming Birds (*Trochilus polytmus*), Jamaica.

50

and black seeds, and an acid pineapple flavour. The avocado pear too was good as a salad; it looked like a pear, only sometimes it was purple as well as green, and had a large seed inside but the white part had the consistency of a very ripe pear without the slightest taste. I used to wander up the hill-paths behind the house in the evening and make friends with the logwood-tree, just then covered with yellow flowers: the anotha with pink or pearl-coloured buds and wonderfully packed crimson seeds in husks like sweet chestnuts wide open. One could hold these prickly shells upside down and shake them and the seeds never shook out, the prickles being curved over their surface, so that they were secured as with a network.

The principal palms on the hills were the cabbage, the young shoot of which is eaten boiled, for which the poor tree is killed; the "maccafoot" and the "groo-groo," whose great seeds take a high polish, and look like onyx stones in a bracelet: the mahogany-cones open in four leaves, and the seeds inside are packed like French bonbons in lace-paper. I was always finding fresh wonders. The sea-cucumber, a gourd which grew near the shore, had the most wonderful mat or skeleton sponge rolled up inside, which the natives used as a scrubbing-brush. The delicious star-apple got ripe, and was filled with blancmange flavoured with black currants.

One Sunday morning I walked up to Craigton, and on to Judge Ker's. I got up my 1800 feet before eight o'clock, and found his worship in an extra scarecrowish costume gardening. He was a very odd man, but was one of the people I liked, so original and honest, it was difficult to listen to his talk without laughing. He lent me his good gray horse, and I rode up to the church, and asked Mr. B. to get me leave to go and stay at Clifton Lodge, which he did. The house belonged to a gentleman who had lost his wife there, and never cared to see it again; he did not let it, but *lent* it for a week at a time to different people, who wanted a dose of cool air, 5000 feet above the sea, beyond the lovely fern walk and in the midst of the finest and oldest coffee-plantations in Jamaica. Opposite was the real Blue Mountain, with clouds rolling up across it as they do in Switzerland. There was a village just below, with a great coffee-growing establishment, and bushes of it for miles on the hillside in

front – all pollards, about four feet high, full of flowers and different coloured berries. It seemed an ill-regulated shrub; its berries had not all the same idea about the time for becoming ripe, and the natives had to humour them and pick continually.

I did one great study in the Fern Walk, sitting in my mackintosh cloak, and bringing it back soaking outside every day. Then one afternoon a dragoon arrived on horseback with a letter asking me for a week to Spanish Town Government House to meet the S.s only. The balls and heavy parties being over I could not resist, though sorry to leave the nice place I was in. I reached Spanish Town in the dark, barely in time for dinner, and enjoyed all the more looking out at my window the next morning on the lovely convent-like garden below, full of the richest trees and plants. When the day's work was over, Gertrude and I used to go and sit there too: I still painting in a corner, she and her brother and friends swinging in hammocks and talking nonsense, till the great heat was over, and we could go for a ride or drive.

I was told it was nonsense keeping the garden-house any longer, as I had so many other houses, so I resolved to go over and give it up. I went by rail to Kingston over a rich plain of grass-land, dotted like an English park with magnificent trees, mostly of the flat-topped rosewood and allied species, passing also some giant cotton-trees, which adapt themselves to the flat land by growing in width rather than height: their buttresses were huge. I got a carriage and drove up to the garden-house, paid off my two old retainers, and packed up my things.

I had one delightful day in the Bog Walk with the M.s. I saw the great aristolochia trailing over the trees, as evil-smelling as its neighbour the portlandia was sweet. Tangles of ferns and orchids on every rock, with the clear river rushing among them, sometimes bounding between huge rocks, sometimes winding along serenely between great plumes of feathery bamboo. It was a glorious four miles of scenery, but too far from Spanish Town to work it comfortably.

Two days afterwards I packed myself and my trunk into a two-seated box on wheels, with a flat waterproof top, and curtains tied up with bits of

A night-flowering crinum lily (*Crinum* sp.) and ferns, in Jamaica.

A Vision of Eden

Foliage, flowers and fruit of the Nutmeg Tree (*Myristica fragrans*) and a Vervain Humming Bird (*Mellisuga minima*), Jamaica. The butterfly is *Battus polydamus*.

string. It had two horses attached to it; one pulled it and one ran beside the one who pulled. The former fell down and broke its knees at the first hill, which taught me a useful moral: never work too hard nor try to do more than your neighbours, or you may break your knees, which is unbecoming even in a horse! Morant Bay was too tempting to pass, especially as it had a good inn kept by Miss Burton, a large black lady with most amiable manners. The house was raised above the village; she gave me a nice corner room with a large tub in it – very acceptable after coming from the mosquitoes of Kingston – and I began a sketch at once of a great cotton-tree half growing in the river, with the blue sea beyond, shaded by palms and bamboos.

After leaving the sea the atmosphere got more and more like a hot fern-house, till we reached Bath, where the inn was kept by a decent kind of white woman. It was really hot and without air; so I worked at home in slight clothing till four o'clock, and then walked up two miles of marvellous wood scenery to the baths, which were slightly sulphurous and very hot and delicious. Two large nutmegs, male and female, grew close to them, with the beautiful outer fruit just opening and showing the nut and the crimson network of mace round it. The flowers are like those of the arbutus. The town of Bath consists of one long street of detached houses, having an avenue down it of alternate cabbage-palms and Otaheite apples. The old botanical garden had long since been left to the care of nature; but to my mind no gardener could have treated it better, for everything grew as it liked, and the ugly formal paths were almost undiscoverable. The most gorgeous trees were tangled up with splendid climbing plants, all seeding and flowering luxuriantly; the yellow fruit of the gamboge strewed the ground under them, and the screw-pine rested on its stilted roots, over which hoya plants were twining, covered with their sweet star-flowers.

I asked why I saw no snakes, and was told they had all gone up into the trees to drink out of the wild bromeliads! Those pretty parasites often held quite a pint of water in the cornucopias which form their centres, as I found to my cost one day when bending one down to look at its flower, and it emptied its contents up my sleeve.

The road eastward was very lovely, making short cuts from one beautiful bay to another, passing many little landlocked harbours of the very deepest blue, with cocoa-nuts fringing the very edge of the sea, and grotesque rocks hollowed out by the waves underneath, hung with leaves of maidenhair a foot long. We rested during the mid-day heat at Manchineal, where I sat in the doorway to draw a palm, and the fattest hostess I ever saw sat beside me, cutting up guavas into a pot to make jelly of, while her little boy of four cracked Palma-Christi berries with his teeth preparatory to making castor-oil. We did not reach Port Antonio till after sunset, so many attractive things had tempted me to linger and sketch. The hotel was full, but they gave me a room out, in the house of a very beautiful brown lady. During the night the rats came and ate holes in my boots, which were very precious and not easily replaced, so I always put them on the top of the water-jug during the rest of my stay on the island.

My next resting-place was at Mr. E.'s, where I stayed a week. His home was perched on a rock like a fortress; one could see miles of cocoa-nuts on one side and of sugar on the other. The mountains came down close behind it, over which twenty miles of rough riding road could take one to Newcastle, and a very beautiful road it must have been. Mr. E. had a large sugar farm, and enough to do – he was out from sunrise to sunset. The sugar-canes grew here magnificently, planted sufficiently wide apart to allow a plough to be worked between the rows. They threw up from fifty to eighty canes in one bunch, and were often fourteen feet high. Rats are their chief enemies, gnawing the cane near the ground so that it falls and dies. A penny was offered for every dead rat, and often 1000 were killed in one week. The governor had introduced the mongoose from India to eat the rats, but they preferred chickens, and rather liked sugar too, so were, on the whole (like most imported creatures), more harmful than beneficial.

One of the amusements at dinner was to play with a kind of small cockchafer, no bigger than an English house-fly, but so strong that he would carry a wine-glass on his back easily across the table. Two or three used to be set to run races in that way, and one of them once carried a small salt-cellar full of salt in the same way. The next house I stopped at was over another bay on a high hill-top, with most exquisite sea and land views over a park-like country, with groups of richest trees and palms; but they blew about too much to paint with comfort.

My host was one of the largest growers and makers of sugar, and managed seven other estates besides his own. One afternoon we climbed the 800 feet of steep park-like road up to Shaw Park, the very gem of all Jamaica, where I was received with the heartiest of welcomes by Mrs. S. and her wild family. The house stood on a wide terrace of smooth green turf; wooded hills rose behind it; real forests of grand timber trees – teak, cedar, fiddle-wood, and astic, cocoa-nuts and cabbage-palms – came close to it; and on one side was a gully with masses of bananas and ferns, and a large fallen tree to act as a bridge over the stream, with a washerwoman in bare legs always ready to hand one across.

The air was always fresh at Shaw Park, but there was little shade just round the house, as the trees had been cut away to make places for drying the allspice – great floors of cement side by side covering as much space as a house. One tree yielded eleven shillings' worth of fruit in good seasons.

One night while at dinner we heard a great screeching of hens and cocks; a black man was sent out to see if it was a snake, and soon returned breathless: "Him bery big yaller one, him wait for Massa Jim, come kill him." We all jumped up in a great commotion. Jim seized a great old sword from the wall, I headed the party, and we found that the niggers had driven the snake up into a tree after it had killed one chicken and nearly caught the old hen. And now the black people were dancing round and round the tree, and singing out: "heh! heh! him bery big," at the tops of their voices to keep him up there. Jim quietly pushed his way through the ring, climbed up the tree, and after a St. George and the dragon fight cut off the snake's head, the big beast hissing and spitting at him to the last. The butler brought the great body in wound round and round a branch six feet long, and as thick as my arm; they said his wife would be sure to come to look for him the next night.

The road along the coast to St. Anne's was shaded by bread-fruit and mammee-trees; the *Broughtonia sanguinea* orchid was hanging like a string of rubies from the rocks among the fresh green ferns. I was sorely tempted to take a small vacant house there called Eden Bower for £3 a month, with endless cocoa-nuts and grass for the horses, and enough allspice to pay my rent (fever also in plenty).

Prudence, however, drove me back to the civilised side of the island over the Monte Diabolo. Ascending by the very ferniest gully I ever saw, where the banana leaves were absolutely un-broken by any wind, we came to a kind of alpine scenery – a wide waving table-land of grass with

Cultivated flowers painted in Jamaica. Below on the left is a white-flowered *Gardenia* and a showy scarlet *Passiflora quadriglandulosa*. Above is a Red Angel's Trumpet (*Brugmansia sanguinea*) and *Broughtonia sanguinea*, with a cluster of Lady-of-the-Night (*Brunfelsia americana*) in the centre. On the right is a leaf of *Caladium*, a spike of Pride of India (*Lagerstroemia speciosa*) and flowers and fruit of the Passion Flower (*Passiflora alata*).

trees dotted about it, oranges, allspice, and different timber trees hung with orchids, but not in flower. They were harvesting the oranges in one place in the usual way when the "Massa" or "Busha" is not by, that is, sitting in groups under the trees and eating them. I stayed a night at Linstead, a pretty village at the head of the famous Bog Walk, and the next day drove all the way through Spanish Town, with its big deserted Queen's House, to Kingston, and climbed the hill to Bermuda Mount to stay with Gertrude S. and her brother the Attorney-General.

It was delightful to be with such people again – people who read and thought, and enjoyed a joke too, and were never idle. We were very happy together; though the summer heat prevented me from working out of doors, I always found abundance of flowers to paint in the cool verandah.

1872. – On the 24th of May Gertrude took me on board the *Cuban* – a roomy ship, with delightful deck cabins, and a jolly captain. On the 16th of June we landed at Liverpool, and two days after I was at home.

CHAPTER IV

Brazil

1872–73

FOR THE NEXT TWO MONTHS I enjoyed the society of my friends in London, and then began to think of carrying out my original plan of going to Brazil, to continue the collection of studies of tropical plants which I had begun in Jamaica.

1872. – I started in the *Neva* Royal Mail Ship on the 9th of August with a letter from Mr. R.G. to the captain. I had a most comfortable cabin, quite a little room, with a square window, and the voyage was most enjoyable. Lisbon was our first halt, which we reached on the 13th at sunset; the entrance to the harbour is striking, with the semi-Moorish tower and convent of Bela in the foreground; on the 19th we stopped to coal at St. Vincent. I did not land on that treeless island, which looked like a great cinder itself.

On the 28th of August 1872 we cast anchor at daylight off Pernambuco, and I saw the long reef with its lighthouse and guardian breakers stretching out between us and the land, and wondered how the crowd of ships with their tall masts ever got into the harbour. Seen through my glass, the buildings of the town looked much like those of any other town, but beyond were endless groves of cocoa-nut-trees, showing clearly in what part of the world we were.

It was Sunday, and the shops were shut with as much rigour as in Glasgow itself. I saw little to buy but parrots, oranges, and bananas; no ladies were about, they were all in church. But though the upper class of women was wanting, there were plenty of negresses in the streets, whose gay-coloured striped shawls hung over their heads and shoulders in the most picturesque folds; and in the suburb gardens we saw grand palms and other tropical plants new to me. The fan-palm of Madagascar was perhaps the most remarkable, with its long oar-like leaves and stalks wonderfully fitted together in the old Grecian plait, each stalk forming a perfect reservoir of pure water, easily tapped from the trunk; thirsty travellers had good reason for naming this palm or strelitzia their friend.

At Bahia we also landed, and after mounting the steep zigzag to the top of the cliff, had another drive into the country, which is wild, hilly, and covered with rich forests. The *Neva* took us in two more days safely into the beautiful Bay of Rio, which certainly is the most lovely sea-scape in the world; even Naples and Palermo must be content to hold a second place to it in point of natural beauty. I know nothing more trying to a shy person than landing for the first time among a strange people and language, I always dread it; so I asked the good Belgian merchant to help me, and he gave me into the care of one of his brothers, who not only landed me in his boat, but put me into a carriage which took me to the Hôtel des Étrangers at Botofogo, on the outside of the town.

I soon felt myself at home in Rio, and in a few days had a large airy room and dressing-room at the top of the hotel, with views from the windows which in every changing mood of the weather were a real pleasure to study; both the Sugar-loaf and Corcovado mountains and part of the bay also were within sight.

The mule-cars passed the door of the hotel every ten minutes, and took me at six o'clock

every day to the famous Botanical Gardens, about four miles off. The gardens of Botofogo were a never-ending delight to me; and, as the good Austrian director allowed me to keep my easel and other things at his house, I felt quite at home there, and for some time worked every day and all day under its shady avenues, only returning at sunset to dine and rest, far too tired to pay evening visits, and thereby disgusted some of my kind friends. Of course my first work was to attempt to make a sketch of the great avenue of royal palms which has been so often described. It is half a mile long at least, and the trees are 100 feet high, though only thirty years old; they greatly resemble the cabbage-palm of the West Indies, though less graceful, having the same great green sheaths to their leaf-stalks, which peel off and drop with the leaves when ripe; about five fell in the year, and each left a distinct ring on the smooth trunk. The base of the trunk was much swollen out, and looked like a giant bulb. This huge avenue looked fine from wherever you saw it (and reminded me of the halls of Karnac). There were grand specimens of other palms in the gardens: a whole row of the curious Screw-Pine, with its stilted roots and male and female trees; rows of camphor-trees, bamboos, the jack-fruit, with its monstrous pumpkin-like fruits hanging close to the rough trunks, and endless other interesting plants and trees.

After a fortnight's daily work there the weather became cloudy, and I brought home flowers or fish to work at, my landlord kindly letting me go with him any morning I liked to the wonderful market, where the oddest fish were to be found, and where boat-loads of oranges were landed and sold all day long on the quay-side.

At Rio I made my first acquaintance with a very common inhabitant of the tropics, a large caterpillar, who built himself first a sort of crinoline of sticks and then covered it with a thick web; this dwelling he carried about with him as a snail does his shell, spinning an outwork of web round a twig of his pet tree, by which his house hung, leaving him free to put out three joints of his

Foliage and flowers of *Strophanthus gratus*, a climbing plant from West Africa, with Royal Palms (*Roystonea regia*) and the Sugarloaf Mountain, Brazil, in the background.

head and neck, and to eat up all the leaves and flowers within his reach; when the branches were bare he spun a bit more web up to a higher twig, bit through the old one, jerked his whole establishment upstairs, and then commenced eating again. He had a kind of elastic portico to his house which closed over his head at the slightest noise, his house shutting up close to it like a telescope; and then when all was quiet again out came his head, down dropped the building, and the gourmand again set himself to the task of continual eating. He ate on for some months incessantly, using his claws to push and pull dainty bits down to him, and shifting his moorings in a most marvellous way. At last the sleep of the chrysalis overtook him, and he finally became a very dowdy moth.

I spent some days in walking and sketching on the hills behind the city; its aqueduct road was a great help to this enjoyment, being cut through the real forest about a thousand feet above the town and sea. In this neighbourhood I saw many curious sights. One day six monkeys with long tails and gray whiskers were chattering in one tree, and allowed me to come up close underneath and watch their games through my opera-glass; the branches they were on were quite as well worth studying as themselves, loaded as they were with creeping-plants and grown over with wild bromeliads, orchids and ferns; these bromeliads had often the most gorgeous scarlet or crimson spikes of flowers. The cecropia or trumpet-tree was always the most conspicuous one in the forest, with its huge white-lined horse-chestnut-shaped leaves, young pink shoots, and hollow stems, in which a lazy kind of ant easily found a ready-made house of many storeys. The most awkward of all animals, the sloth, also spent his dull life on the branches, slowly eating up the young shoots and hugging them with his hooked feet, preferring to hang and sleep head downwards. Some of the acacia-trees grow in tufts on tall slender stems, and seem to mimic the tree-ferns with their long feathery fronds, whose stems were often twenty to thirty feet high.

Of course (again), like all other visitors to Rio, I walked up to the top of the Corcovado and looked down on the clouds and peeps of blue sea and mountains seen occasionally through them, and on the splendid yellow and white amaryllis

clinging to the inaccessible crannies of the rock; the whole way was a series of wonders and endless beauties.

On that expedition I met, for the first time, Mr. Gordon and his daughter, who asked me to come and see them in Minas Geraes, to which they were returning in about three weeks. I liked their looks and manner of asking me, and it seemed a grand opportunity of seeing something of the country, so I said I would come for a fortnight, at which they laughed, and with reason, for I stayed eight months! I first visited the Island of Pakita then stayed at Tignea.

On the 25th of October I sent down my three portmanteaus in a return-cart and followed myself the next day, in pouring rain, to Rio. After some necessary shopping and other business, I crossed the bay and its lovely islands for Mawa, where a train was waiting to take us over the marsh to the foot of the Petropolis hills. At last we reached a more healthy-looking region, and stopped at Reiz da Serra, where I was put into a carriage with three Brazilians and conveyed up the ten miles of zigzag road, dragged by four mules, who kept up a continual trot, the rise of 3000 feet being well graduated. Two more miles at full gallop down hill took us to Petropolis, and I was soon in Mr. Miles's comfortable hotel, and again among friends, with whom I had a merry English dinner. Then came two days of rain and cold and loneliness, in which I worked and walked and soaked and froze, and came to the conclusion Petropolis was an odious place, a bad imitation of a second-class German watering-place. I was glad to see the Gordons arrive, and to hear them say they had taken their and my places in the coach for Juiz de Fora the next morning. It rained all night, and was still raining when we packed ourselves into the coach at six on the morning of the 28th of October, and four splendid mules, after their usual resistance, started suddenly at full gallop with the swinging, rattling old vehicle. A violent jerk brought us to the door of the other inn, and there our fourth place was filled up by a very important person in these pages, Antonio Marcus, commonly called the Baron of Morro la Gloria, who had been for forty years in the service of St João del Rey Mining Company, to whose mines I was going.

Such scenery! High trees draped with bougainvillea to the very tops, bushes of the same nearer the ground reminding one of the great rhododendrons in our own shrubberies in May at home, and of much the same colour, though occasionally paler and pinker. There were orange-flowered cassia-trees (whose leaves fold close together at night like the sensitive plant) and scarlet erythrinas looking like gems among the masses of rich green; exquisite peeps of the river, winding below its woody banks or rushing among great stones and rocks, came upon us, and were gone again with tantalising rapidity. The wild agaves too were very odd: having had their poor centre shoots twisted out, the sap accumulated in the hollow, and a wine or spirit was made from it; the wretched wounded things, sending up dwarfish flowers and prickly shoots from their other joints, formed a strange disagreeable-looking bush, several of which made a most efficient hedge. Under each of these flowers a bulb formed, which when ripe dropped and rooted itself, thus replacing the parent whose life ended at its birth. We stopped to dine at Entre Rios; here we came to the Don Pedro railway, and the real traffic of our road began. There was no other way of reaching the rich province of Minas, or of obtaining its minerals, coffee, sugar, or cotton; so from this point we passed a continual stream of mules or waggons till we got to Juiz de Fora and its most comfortable hotel.

Every one said the road to Minas was impassable from the late heavy rains. We heard of mules being smothered in the mud, a woman killed in it, etc.; but the more I heard the more I determined to see my friends safely through, if they were willing to be burdened with me; besides, people had said in Rio I should never really go, some had done their best to keep me from going, and one Scotchman had said I should "not find to paint any in Minas!"

The first loading of thirty-seven mules is not done in an hour; everything must be weighed and strengthened and hung with stout bands of cowhide, balanced well, or the mules will suffer. When once they are well loaded the things are numbered, and the operation on subsequent mornings becomes a much easier and quicker affair. All these arrangements were our Baron's glory; he had to think and be responsible for every little item, and made as much fuss as he possibly

Flowers cultivated in the Botanic Garden, Rio de Janeiro, Brazil. Red and white Indian water lilies, with the large flowers, probably of a *Solandra* species, and the crimson flowers of an Australian one-sided bottlebrush (*Calothamnus* sp.).

could, getting in and out of a score of terrible rages before midday. When the rain left off, his temper also cleared, and we finally started, forming a party which would not have shone in Hyde Park, but was admirably adapted for riding through Brazil in the wet season.

First went the loaded mules with their bare-legged black drivers, then the Baron in the shabbiest of straw hats, any quantity of worsted comforters, and brown coat and gaiters. Mr. G. on his noble gray mule, his daughter on her pretty little horse, and myself on Mueda, the steadiest and most calculating of mules. The road was one constant succession of holes and traps and pies of mud, often above the mules' knees, often worn by constant traffic into ridges like a ploughed field, through which the tired quadrupeds had to wade, or drag their feet from furrow to furrow of the sticky, soft, clogging mud. Every traveller we met delighted in magnifying the horrors they had passed, and said that as the rain had continued it was utterly impossible for us to go on; and one party which had started the day before were actually coming back in despair. Our progress through all of this was slow; we were obliged to stop after only 3½ leagues of it, and put up for the night, while Mr. G. sent on a note to the chief engineer of the province to ask his help. An answer came the next morning, begging us not to start too early; he had set fifty men to work, and hoped to make the road passable by noon, which gave us time to enjoy and examine our present quarters. There was a particularly bad place opposite our door; it probably had been particularly bad for years, and would be the same for years to come, it having apparently never come into the head of the landlord to mend it. Perhaps he thought it stopped people and brought custom to his house, as they were literally unable to pass his door. One by one we saw the poor mules go flop into the liquid mud-hole, have their loads transferred to men's heads, and themselves lifted out by tail and head, the lifters often replacing them in the hole during the process. We, however, all got safely over, and were soon met by "Beesmark himself," as our Baron called the great Prussian engineer, a large man with a magnificent white beard and tall horse, which I believe was once of the same pure colour. After many compliments and hearty greetings he took the lead, and we rode round the valley by the steep hillsides, so as to avoid the muddy road and marsh, now powdered with lovely masses of the *Francisea*, with its blue and white blooms.

Our next night's quarters were worse than the first; for the landlord had not been out of his house for a month, and had not even a sack of corn for our poor tired beasts; but the night after that we passed in a fazenda or farmhouse, with a beautiful green grassy hill behind it, on which the animals did enjoy themselves, rolling over and over, cleaning their coats, and eating any quantity of delicious *capim* grass. Near here I first saw the araucaria-trees (*A. braziliensis*) in abundance; it is the most valuable timber of these parts, and goes by the name of "pine." The heart of it is very hard and coloured like mahogany; from this all sorts of fine carvings can be made; the outer wood is coloured like the common fir. This tree has three distinct ages and characters of form: in the first it

looks a perfect cone; in the second a barrel with flat top, getting always flatter as the lower branches drop off, till in its last stage none but those turning up are left, and it looks at a distance like a stick with a saucer balanced on the top. During the first period the branches are more covered with green; but as it grows older only the ends are furnished with bunches of knife-like leaves, and the extremities alone are a bright fresh green, looking like stars in the distance among the bare branches and duller old leaves. Its large cone is wonderfully packed with great wedge-shaped nuts, which are very good to eat when roasted. These curious trees seldom grow lower than 3000 feet above the sea.

After crossing the grand pass of Mantiqueira we changed the general character of vegetation. I saw there masses of the creeping bamboo, so solid in its greenery that it might have been almost mowed with a scythe; also the Taquâra bamboo hanging in exquisite curves, with wheels of delicate green round its slender stems, reminding me of magnified mares' tails, and forming arches of 12 to 20 feet in span. Every bit of the way was interesting and beautiful; I never found the dreary monotony Rio friends had talked about. Every now and then we came to bits of cultivation, green hills, and garden grounds. Once I saw a spider as big as a small sparrow with velvety paws; and everywhere were marvellous webs and nests. How could such a land be dull? Then we crossed high table-lands which seemed quite colonised by the "Jean de Barbe" bird; every tree was full of their nests – curious buildings of red clay as big as my head, divided into two apartments.

After a long day's ride over these glaring plains, still sticky and slippery with mud, though the hot sun was shining on it, we were glad to find really comfortable night-quarters in the house of a gentleman who prides himself on producing the best cigarettes in Brazil. At this point in our journey Mr. G.'s carriage met us. The sunshine continued as we rode on over the high country to Barbacena, the chief town of this district, beautifully situated on a hill about 4000 feet above the sea, with fine araucaria and other trees shading

The famous avenue of Royal Palms (*Roystonea regia*) at Botofogo, Brazil.

A palm in the embrace of the roots of a strangler fig tree; after many years the latter forms a complete sheath around the palm trunk, which finally disintegrates leaving the fig standing alone. Nests of the Sociable Oriole (*Ostinops* or *Ocyalas*) are hanging from the tree.

the pigs could not get past him into the house, so why should we mind either of them? Our next quarters made up for Gama; for they were in a friend's house, with a kind Brazilian lady and her children. Mary had a threatening of diphtheria, and longed for home and her mother's care; so we toiled up and down the high ridge of Morro Preto, whose white sharp rocks stuck up like bleached bones, and whose cracks were filled with the brightest red, purple, or yellow earths.

It was most tantalising to pass so many wonders, but time was precious and my friend was suffering, and our next night behind a curtained alcove in an extremely draughty room after a good day's soaking did not improve her. The third morning found her voiceless, but she was determined to get home that night, though it was a full forty miles' ride; so on we came, and she bore it bravely. Suddenly a violent discharge of rockets in front warned Mr. G. he was coming among friends, and we stopped to breakfast at the house of a black man, whose late master had left him his freedom as well as house and property. The journey was a weary one; for we were all anxious about her who was generally the life of our party, and when we reached the bridge over the deep river-bed where we were to change mules I thought she would have been suffocated. Soon, however, the hill of Morro Velho came in sight, and, though still far off, her spirits rose and her troubles grew less in proportion as the distance shortened. A fearful storm came on, and our waterproofs were of real use, and brought us in a comparatively dry state to the house of a very remarkable old lady, Dona Florisabella of Santa Rita, who hugged us all round in the heartiest way, and then led us up by a rough ladder to a set of handsome rooms, which had been frescoed in a most gaudy and reckless manner with every bright tint of the rainbow.

It was no easy task to get away from this hospitable lady, but at last we started, and about a mile farther crossed the great bridge over the river, and were on "The Company's" property. The Baron was low-spirited, for he was no longer our leader, and his work was over. Mr. G. and I led the way and jogged over the muddy road up hill and down to the village of Congonhas, when the rockets and firing and hand-clasping began in good earnest, amid torrents of rain. At last we

its garden slopes. Two tall churches made a finer show in the distance than they did near.

The flowers on these high campos were lovely – campanulas of different tints, peas, mallows, ipomoeas creeping flat on the ground, some with the most beautiful velvety stalks and leaves; many small tigridias, iris, and gladioli, besides all sorts of sweet herbs. There are many peculiar trees and scrubby bushes with brown or white linings to their leaves, and the stems powdered over with the same tints. I have never seen these elsewhere. Gama was said to be the very worst house on the road, and it certainly was not what the Yankee's call "handsome quarters." An idiot sat on the doorstep, pigs wallowed in the mud beyond; but the idiot was said "not to be often dangerous," and were stopped by the band awaiting us, and had to tramp solemnly behind it into the grounds of the Casa Grande – a mass of close-packed dripping umbrellas and damp bodies; and before I knew where I was I found myself dismounted, and hugged and welcomed by one of the best and kindest women I ever met in all the wide world, and called "dearie" in a sweet Scotch voice; no wonder Mary longed to be at home! And I felt that I was right and the Rio people wrong about coming to Morro Velho, and the only drawbacks to the journey left were blistered lips and slightly browned hands.

The Casa Grande of Morro Velho was indeed a rare home for an artist to settle in, and I soon fell into a regular and very pleasant routine of life. I had the cheeriest and most airy of little rooms next my friends, with a large window opening on to the light verandah, in which people were continually coming and going and lingering to gossip. Beyond that was the garden, full of sweetest flowers; a large *Magnolia grandiflora* tree loaded with blossoms within smelling distance; around it masses of roses, carnations, gardenias (never out of flower), bauhinias of every tint (the delight of humming-birds and butterflies), heliotropes grown into standard trees, and covered with sweet bloom, besides great bushes of poinsettia with scarlet stars a foot across; beyond these were bananas, palms, and other trees, and the wooded hillsides and peeps of the old works and stream in the valley below.

I had delightful rambles and always found new wonders on every expedition. Just below the flower-garden was a perfect temple of bananas, roofed with their spreading cool green leaves, which formed an exquisite picture. Sometimes a ray of sunlight would slant in through some chink, and illuminate one of the red-purple banana flowers hanging down from its slender stem, making it look like an enchanted lamp of red flame. Masses of the large wild white ginger flowers were on the bank beyond this temple, and scented the whole air. Farther down the steep path were masses of sensitive plants covering the bank with the brightest of green velvet and delicate lilac buttons. I never could resist passing the handle of my net over this, when instantly the whole bank became of a dull, dead, earthy tint, and only the dry twigs and stalks of the plants were visible, with

their shrinking branchlets starting from them at most acute angles. Below this there were two or three old gray trees, on whose trunks or roots I never failed to find some new wonders of cocoons or larvae, or odd spider's web, green, gold, or silver, as they glittered in the bright morning sun, often spangled with diamond dew.

The curious grass which bears the gray berries called "Job's tears" was also a handsome plant, and abounded here. Beyond the bridge was the kitchen-garden, in which several superannuated black people did as little as possible from sunrise to sunset, at certain very frequent intervals leaving off to light a fire and cook for themselves various sorts of savoury decoctions in an iron pot, taking as long as possible to eat it, after which exertion of course they required rest.

At Morro Velho every man has his garden – such gardens! With running water passing above them so that they could irrigate to any extent, and full of the richest fruit and vegetables. The leaves of their *Caladium esculentum* were often nearly two feet long. At noonday the beautiful banana-leaves lose all their fresh shining greenness, and shut themselves up tight like sheets of folded letter-paper, so as to keep their moisture in, and appear mere knife-like edges to the sun's scorching rays; as it sinks lower they again spread out ready to collect the evening dews. The blacks devote all these garden treasures to their pigs, which they fatten up till they are worthy of Smithfield, with almost invisible necks, little snouts, and short legs.

In a weedy garden near was a humming-bird's nest, hanging to a single leaf of bamboo by a rope of twisted spider's web three or four inches long, swinging with every breath of air. These wee birds generally build in this way about Morro Velho, often hanging their nests over the running water, on the ends of fern-fronds or on the blades of grass, where the eggs are safe from the attacks of snakes or lizards; but they always choose places with some protection above from rain or sun. They sit twice in the year, never laying more than two eggs, which are always white, and about the size of small Scotch sugar-plums. Some of these humming-birds were quite sociable. One pair had come regularly twice a year to one of the outhouses of the Casa Grande for many years, apparently using and repairing the same nest, which hung by a tiny rope from the ceiling.

A Vision of Eden

About January the heat became more oppressive – 86° was the average, though it was often 91° in the shade – but the nights were always cool enough for sleep at Morro Velho, which is about 3000 feet above the sea, and I was never uncomfortably hot. The Gordons, however, who had lived sixteen years in the climate, longed for a change; so they determined to go to pay a long-promised visit to Mr. R. at Cata Branca, taking a young Scotch lady who had been spending Christmas with them, and myself.

It was a beautiful day's ride of about 26 miles, the road winding for the greater part of the way along the high banks overlooking the Rio das Velhas, which eventually runs into the Rio San Francesco, and enters the sea above Bahia. In the fresh clearings I saw many new and gorgeous flowers, as well as some old friends, including the graceful amaranth plant of North Italy, with which the wine of Padua and Verona is coloured. How did it get to the two places so far apart? I longed more and more for some intelligent botanical companion to answer my many questions.

We rested a while at a collection of huts that have been put up for the workpeople near some fine falls of the river, and the Head Man there told us of one curious fish he had caught, which seemed to have a sort of inner mouth, which it sent out like a net to catch small fish or flies with. He showed us a rough drawing he had made, and was very positive about the story, which is not more difficult to believe than many other well-proved wonders of nature. After leaving this settlement, we mounted up bare hillsides another 1000 feet, and came to the green plateau of Cata Branca, with its groups of iron-rocks, piled most fantastically like obelisks or Druid stones standing on end, dry and hard, and so full of metal that the compass does not know where to point. Amid these rocks grow the rarest plants – orchids, vellozias, gum-trees, gesnerias, and many others as yet perhaps unnamed. One of these bore a delicate bloom, – *Macrosiphonia longiflora* (No. 67 in my Catalogue at Kew), – like a giant white primrose of rice-paper with a throat three inches long; it was mounted on a slender stalk, and had leaves of white plush like our mullein, and a most delicious scent of cloves. Another was a gorgeous orange thistle with velvety purple leaves. I was getting wild with my longing to dismount and examine these, when we met our kind host Mr. R. coming out to meet us, and in another half-hour we were in his pretty cottage, where he had been living for the last two years watching a dying mine, in almost perfect solitude, expecting to be released any moment.

The summer of St. Veronica was endless that year, and we had the most glorious weather. The air was much fresher on the height and did us all good. Every day's ramble showed me fresh wonders. One morning we spent on the actual peak, which rises a perfect obelisk of rock 5000 feet above the sea. Some of the more adventurous of our party mounted to the very top.

At last Mr. G. came to fetch us, and on the Sunday before we left he read the Service, three Cornish miners coming up from below to assist at it.

About the end of March we all started up the hills with bag and baggage, crossing over a shoulder of the Coral mountains, and on to Sabara, the chief town of the district, where we took coffee at the house of "a most respectable brown woman," who hugged all my friends most warmly. Our road followed the banks of the river for some way; sometimes along the low banks amongst reeds and bushes of *Franciscea*; often over the higher sierras, among strange scraggy trees, which were covered with more flowers than leaves. On one especially the white lily-like flowers were very fascinating. The ipomoeas and bignonias were in great variety. It was a perfect fairyland. The great blue and opal Morpho butterflies came flopping their wide wings down the narrow lanes close over our heads, moving slowly and with a kind of see-saw motion, so as to let the light catch their glorious metallic colours, entirely perplexing any holder of nets. Gorgeous flowers grew close, but just out of reach, and every now and then I caught sight of some tiny nest, hanging inside a sheltering and prickly screen of brambles. All these wonders seeming to taunt us mortals for trespassing on fairies' grounds, and to tell us they were unapproachable. At last we left the forest, and the real climb began amidst rocks

Flowers of a coral tree (*Erythrina* sp.), Brazil.

68

grown over with everlasting peas, large, filmy, and blue haresfoot ferns, orchids, and on the top grand bushes of a large pleroma with lilac flowers and red buds like the gum-cistus, and beds of the wild strawberry, which some Italian monk had introduced years ago.

The Baron took charge of the two girls and myself over the hills; and at the edge of the Rossa Grande property its superintendent met us, showed us his trim little mine and big wheels, and gave us luncheon, then took us up the hill to admire the view, and accompanied us through two leagues of real virgin forest, the finest I had yet seen, to the old Casa Grande of Gongo – a huge half-ruined house which had originally belonged to some noble family. To this old deserted place Mr. D. had sent furniture, food, and slaves, and persuaded his mining captain to let his pretty little wife go and keep house for us, taking Baby Johnnie to amuse her and us too. She soon made us at home. After tea we played a game of whist in the ghost-like old hall with its heavy wainscoted cupboards, and great gilt hooks from which the mirrors and chandeliers had formerly hung, and on which a late superintendent had once committed suicide.

The next morning our Baron, who had begun life as a blacksmith, went round and mended the locks of all the doors we were likely to use, and our party dispersed, leaving me to enjoy a fortnight of perfect quiet in the great empty house and rich forest scenery, with Mrs. S. and her baby boy to keep me company; to her it was an agreeable change, to me the thing of all others I had longed for. I used to start every morning on my mule for some choice spot in the forest, where I sat down and worked with my brush for some hours, first in one spot and then in another, returning in time for a good wash before dinner. I used generally to roam out before breakfast for an hour or two, when the ground was soaked with heavy dew, and the butterflies were still asleep beneath the sheltering leaves. The birds got up earlier, and the Alma de Gato used to follow me from bush to bush, apparently desirous of knowing what I was after, and as curious about my affairs as I was about his. He was a large brown bird like a cuckoo, with

Aristolochia brasiliensis, a Brazilian climber that has an unpleasant smell.

white tips to his long tail, and was said to see better by night than by day, when he becomes stupidly tame and sociable, and might even be caught with the hand.

One had always been told that flowers were rare in this forest scenery, but I found a great many, and some of them most contradictory ones. There was a coarse marigold-looking bloom with the sweetest scent of vanilla, and a large purple-bell bignonia creeper with the strongest smell of garlic. A lovely velvet-leaved ipomoea with large white blossom and dark eye, and a perfectly exquisite rose-coloured bignonia bush were very common. Large-leaved dracaenas were also in flower, mingled with feathery fern-trees.

A messenger at last recalled us to Morro Velho. Visitors had arrived, and Mrs. Gordon wanted us to help in entertaining them; so we obeyed at once, stopping by the way to breakfast with the Baron's family, to his great delight.

A fortnight's work at home was very pleasant, with kind friends, occasional visitors coming to relieve the monotony of a family party. Amongst others, a gentleman came, a real genius, who combined the accomplishments of making false teeth and tuning pianos; he also excelled as a dog-doctor and maker of guitars. Once we made a picnic to the top of the Coral mountain, 2000 feet above us.

At last Mr. Gordon said he would start on the 21st of May for his long-talked-of holiday-journey to the Caves of Corvelho, and arranged to take his daughter and myself and an English gentleman. We were to have started at daylight, but many things came to detain our leader, so that we really did not get off till the full heat of midday. We passed the night at a lonely farm-house – the roughest of quarters, where we were glad to camp round a wood fire on the mud floor, sitting on our wraps. The next day we rode on through a wooded country, till at sunset we came in sight of the Lago Santo, a shallow sheet of water getting gradually filled up, and I could hardly see the likeness our old Swiss friend found between it and the Lake of Geneva. It was certainly long since he had seen the latter, he said.

An air of indescribable dulness seemed to hang over it and the poor straggling village on its banks, and we wondered more and more what the charm was which had kept the famous Danish naturalist

A Vision of Eden

Dr. Lund here for more than forty years. He was now nearly eighty years old, and had made several collections of natural curiosities and plants, which had gone to Copenhagen. He had corresponded with many of the scientific men of Europe, but always lived entirely alone, and since the death of his secretary he seldom had an educated man to speak to. Once the Danish Government sent a man-of-war to Rio to fetch the doctor home, and he rode as far as Morro Velho, then lost courage, and returned to his dear lake. In former days he used to pass the heat of the day in a room he had built and fitted up as a laboratory over his boat-house on the lake; but now his habits were those of an invalid, and he seldom went outside his garden, and never left his room till after midday, when he liked to sit in his arbour and talk, which he did well in many languages. His garden was full of rare plants and curiosities, collected and planted by himself. The trunk of one large date-palm was covered with a mass of lilac Laelia flowers, and a beautiful night-blooming cactus hung in great

Below: Tree ferns and climbing bamboos in the Gongo Forest, Brazil.

festoons from another tree, or climbed against a wall like a giant centipede, throwing out its feet or roots on each side to cling on by – it seemed to change its whole character by force of circumstances. I had made a painting during the morning of a rare blue pontederia which the doctor had persuaded with considerable difficulty to grow on his lake, and he was much delighted with it, and declared I was "one very great wonder" to have done it. The *Philodendron lundii* was another of his most curious plants, a sort of great cut-leaved and climbing tree-arum, whose leaves are almost as good as a sundial, showing by their temperature the time of day.

Thirty years before our visit Dr. Lund had discovered the stalactite caves of Corvelho, and, as we were now on our way to them, was much interested in giving us directions how to find parts of them unknown to any but himself. Few persons had been far in since he first found them. He told us of the large apes, lizards, snakes, and other antediluvian beasts whose bones he had found there, as well as those of men with retreating foreheads, whose teeth showed they lived on unground corn and nuts. On the way to the caves we stopped on our way to explore some smaller ones, the entrances of which were beautifully draped with cacti and other parasites and creepers, from under which flew a troop of beautiful white owls. A league beyond this was the great cave we had come so far to see, whose entrance was reached by a steep climb of a hundred yards from the stream below. It was quite dry, but the steps or terraces which marked the different water levels were edged and banked up as regularly as the fountains of old Rome, and the grand hall at the entrance was like a bit of fairyland. Great masses of stalactite stood up from the ground, or hung from the roof, tinted with delicate blues and greens and creamy whites. Within the cave our scanty supply of light did little towards showing the endless halls and passages, and as Mr. Gordon was bent on

Opposite: The delicate blooms of the Flannel Flower (*Macrosiphonia longiflora*), so called for the woolly covering to parts of the plant, and butterflies – *Catagramma* species on the left, and *Callicore clymena* species on the right – painted at Casa Branca, Brazil.

A Vision of Eden

Above: Scene in Dr. Lund's garden, Brazil. The large trunk in the foreground is covered with a cactus (*Cereus* sp.), a large aroid (*Philodendron* sp.) and orchids. On the left is a palm (*Acrocomia* sp.) with a mass of *Cattleya* orchids on its lower trunk.

Opposite: Poinsettia or Christmas Flower (*Euphorbia pulcherrima*) painted at Morro Velho, Brazil. This plant was cultivated as an ornamental during Victorian times; as today, it was popular for the red bracts that surround the insignificant flowers.

making a minute measurement of them all, the progress was slow. The width of one of the great elliptical roofs was over fifty French metres, and it had no pillars or supports, only the elegant pendent stalactites hanging in groups. We found abundant footmarks of ounces, gambats, and pacas, but only one place where bones were buried in the stalactite, and they were too much broken to be worth moving. Four long days we passed in this wonderful cave.

Our rides home were by the light of a very young moon and the fireflies. We tried to keep the cloud of white dust kicked up by the mule in front at a sufficient distance to guide us, and yet not get into our eyes. It was tedious work, and at the end of our journey we found little rest. So after the second day underground we determined to try another Fazenda. Mounting the wooded heights above the cave, then over open downs dotted with silk-cotton and other trees, one caught fine distant views of the Diamantina mountains, dark purple in the sunset, which was a rare gold and vermilion tint that night. The show was scarcely over when we reached "Once." The name sounded ominous, but the people were kind, and though somewhat astonished at such an invasion, received us hospitably as we rode into their enclosure.

A most jolly fat lady came out, lightly attired in the usual embroidered chemise and a red petticoat. She took Mary and myself at once to see our "rooms" – a large barn with many tiles wanting in its roof, and well ventilated walls, half filled with looms. How cold it was! How the wind whistled through the holes in the wall close to us! Mary said she should die if she stayed a second night, but she did not; for we had another long day in the cave and another night in the barn, and then rode over the windy sierras back to Cedros.

Our host rode on with us and lost his way before we had gone three miles, though we were bound for the principal town of that district! We stopped at a really decent little inn kept by an old black woman named Donna Anna. Our room was full of sacks of grain and bales of groceries; but the beds were covered with gorgeous quilts, the linen dazzlingly white, edged with fine lace and knotted fringe, also made by hand. We had even a looking-glass, and a basin and jug; but these luxuries had to go the round of the guests from room to room, including a strolling photo-

74

grapher, whose chemicals occasionally sent the water in rather black.

Another cold day's ride brought us to a large farmhouse belonging to a very remarkable family, who would have made their fortunes at fairs. The farmer himself was perfectly round, with a bullet head and face, all over which grew hair and beard apparently cut with his wife's bluntest pair of scissors as close as he could with his left hand. His fat wife had a thick black beard and moustache (uncut), her grandmother the same in gray. The children were all perfectly round, like their fascinating parents, but as yet beardless. We had a grand but greasy dinner; the table quite groaned beneath the quantity of heavy dishes on it. A small pig cooked whole, and considerably over the usual size for making such a barbarous exhibition of itself, was among the dainties.

It was a relief the next morning to hear the tremendous voice of our friend the Baron R. de V. before daylight, shouting to the gentlemen in the adjoining room. He seemed to bring a more genial world nearer to us. He had ridden over in a wonderful peaked woollen hood to make sure we did not pass his house without going in. Now his house was not in our road at all; but it was impossible to refuse our friend's positive determination to take us there, and we resigned ourselves to his will. It was very refreshing to sit a while with the quiet Baronessa and her silver coffee-tray and Minton cups, all so bright and clean; but after a rest and a chat we went on to Dr. Lund's again, and were obliged to stop another night at the grocer's, as the old gentleman never came out of his shell before midday. After a while we rode on a little farther to the house of our old Swiss friend, which he had turned outside in for us. He would let no one wait on us but himself, had cut down twenty dwarf palm-trees to make one dish of cabbage for us, and hung up various branches and bright flowers about the verandah; a decoration no native would think of.

We had but a short journey on to the "City" of Santa Lucia, a most picturesquely situated village on the top of a hill, looking over a long stretch of the winding Rio das Velhas, which again reminded me of the Tweed, and except for a few palm-trees looked not a bit more tropical; while the churches, with their metal pepper-pot towers, and the tiled roofs of the one-storeyed houses,

View from the Sierra of Petropolis, Brazil, with the Bay of Rio, its islands and the Sugarloaf Mountain in the distance.

suggested Hungary. From Santa Lucia our way was hot and dusty as we crept round the shoulder of the Piedade mountain, and came at last to a ridge from whence we looked down on the pretty town of Sabara, descending by a road so steep that walking was almost a necessity, after which we were glad to escape from the glare into the shelter of Donna Anna's roof, where we lingered till the cool evening and rode home by moonlight; what luxury "home" was after such a three weeks of wandering!

On the 2nd of July I saw the last of dear old Morro Velho, and accompanied Mr. and Mrs. G.

back to Rossa Grande, and on to Cocaes. The forest of Gongo had lost much of its beauty during this cold dry season; more trees had lost their leaves than I expected in a tropical country, and flowers were quite rare. I was rather glad of this, as it made me regret less that I was leaving so lovely a country, and I took away the hope of seeing my kind friends again in England; but in spite of this it was hard to say good-bye to dear Mrs. Gordon at Cocaes. Mr. G. was to go down with me to Caraca, so down we went to a bare hill country, and leaving the village of St. John to our right, soon came to the bridge and ravine of Caite. It was a

fête day, and everybody was on the road dressed in their best.

The neighbourhood abounded in rare orchids and other plants, but the rain never ceased to pour, and at this time of year it generally did pour on these mountains; so there was little use in staying. I said good-bye to Mr. Gordon, who had loaded me with such continual kindness and hospitality for the last eight months. He now returned to Cocaes, while I rode after the Baron in the opposite direction.

We crossed the high boggy watershed, every pool and river being bordered with a curious

A Vision of Eden

dwarf bamboo peculiar to this mountain, more like young cypresses than canes; and the rocks were everywhere covered with rare orchids. The wild bromeliads were glorious; I saw acres of one I had given ten shillings for not many years ago at Henderson's – a nidularia with deep carmine nest and turquoise flowers in the centre. After much sliding, tumbling, and slipping, we arrived at Senor Antonio de Sonlea's, and were received most kindly by him and his young wife. Before eight the next morning we were riding over the smaller spurs and still under the wet clouds of Caraca, now and then getting a good shower-bath from some overhanging curl of bamboo or green tangle as we passed. Everything was dripping with moisture; how lovely those wet mornings were! And the huge spiders' webs all strung with crystal beads, so strong that they seemed to cut one's face riding through them.

My last night on this journey was an unquiet one, in another solitary house near the new railway works. It was Sunday, and half-drunken navvies came and thumped at the door all night. The next morning, after squeezing the good old Baron's hand for the last time with real regret, I packed myself into the crowded coach and was whirled away towards Rio.

The distant Organ Mountains peeped at us over the ends of the green valleys, and I again thought nothing in the world could be lovelier than that marvellous road; and then what a welcome the kind M.s gave me, and what a cosy little room in their house at Petropolis! It was rather pleasant too to see my old box again and its contents. Of what priceless value those shoes and stockings and paints seemed to me! And how I longed for them! I had intended starting for Para in a week, but was persuaded to give it up, as the yellow fever was still lingering all along the coast; and I had a longing first for rest in my pleasant, comfortable quarters, and then still more for a sight of home, friends, and books again.

Meanwhile I made two visits to Rio, the chief object of which was to see the Emperor, to whom I had a letter from my father's old friend Sir Edward Sabine. The Emperor is a man who would be worth some trouble to know, even if he were the poorest of private gentlemen; he is eminently a gentleman, and full of information and general knowledge on all subjects. He lives more the life of a student than that to which ordinary princes condemn themselves. He gives no public entertainment, but on certain days he and the Empress will receive the poorest of their subjects who like to take their complaints to them. He kindly gave me a special appointment in the morning, and spent more than an hour examining my paintings and talking them over, telling me the names and qualities of different plants which I did not know myself. He then took the whole mass (no small weight) in his arms, and carried them in to show the Empress, telling me to follow. She was also very kind, with a sweet, gentle manner, and both had learned since their journey to Europe (of which they never tired of talking) to shake hands in the English manner. They had both prematurely white hair, brought on by the trouble of losing their daughter and the miserable war in Paraguay. On my second visit to the palace the Emperor was good enough to show me his museum, in which there is a magnificent collection of minerals. He took especial delight in showing me the specimens of coal from the province of Rio Grande do Sul, which promise to be a source of great riches to the country if his schemes of facilitating the transportation can be carried out. At present, though the coal itself is close to the surface of the ground, there are so many transhipments necessary in bringing it to Rio that it is cheaper to bring it from England or the States. I have not the slightest knowledge of mineralogy, but I blacked the ends of my fingers with a wise air, and agreed heartily with the Emperor's opinion, that if the precious stuff could be brought into consumption cheaply, it would be of more use to Brazil than all the diamonds of Diamantina.

Petropolis seemed full of idle people and gossip, and it was thought rather shocking and dangerous for me to wander over the hills alone; wild stories were told of runaway slaves, etc. I felt out of place there, and got more and more home-sick, but determined to have at least a glimpse of the Organ Mountains before I went. I was told the way was most difficult, and even dangerous; neither mules nor guide could be got. Still I persevered, and

Flowers of Angel's Trumpet (*Brugmansia arborea*) and humming birds (*Colubri* sp.), Brazil.

finally heard of a mason at Petropolis who knew the way and would like a change of air and a holiday, but he could only spare four days.

We arrived at Theresopolis by two o'clock, went on for another two leagues, and put up at a quaint and lonely house on the sierra. The boulders there had fallen all round it; they propped it up, and seemed to rest on its roof, and the stables were built under one huge hanging boulder. Great trees and all sorts of rich vegetation grew over and round these big blocks of granite. Beyond all were the most splendid distant views of Rio Bay and its mountains, and over our heads strange obelisks of granite. It was a spot for an artist to spend a life in.

Did I not paint? – and wander and wonder at everything? Every rock bore a botanical collection fit to furnish any hot-house in England. Then there was a real Italian vine pergola leading down through the banana trees to the spring, with picturesque figures continually fetching water from it, and troops of mules, goats, cows, and sheep always moving about; for the grass had failed in most parts of the mountains this year, but was unusually abundant here. I found it hard to leave the next day, and lingered over my work till nearly noon, when a gentleman came down the hill leading his horse, and spoke to me about the view I was taking, then went on and spoke to my guide, arranging with him that as the inn of the place where we were to stop the night was bad, he

should take me to his house, writing at the same time a few lines to his wife, to take with us and explain who we were.

The sky was still red when we reached the little town we were to stop at, and inquired for the address our friend had given us. His young wife would not let us in till she had held a long conversation with us from an upper window, which ended in a good deal of laughing on both sides, she thinking she could talk English, and I Portuguese, and each of us thinking the other talked her own native tongue. After a time her husband came home. I found he was the chief of the police of that province – an educated man of good family. He was extremely curious to know why I was travelling alone, and painting. Did the Government pay my expenses? I certainly could not pay them myself, I was too shabbily dressed for that! I told him when I got home I hoped to paint a picture of the Organ Mountains, and to sell it for so much money that it would pay all my expenses; then at last he understood what I travelled for, for is not money the end of all things?

A few more hours of swamp and a most roundabout road brought us to the foot of the Petropolis sierra, up which I rode, though in time for the train of Passengers from Rio. In three days more I was steaming towards England, and we landed at Southampton on the 14th of September.

CHAPTER V

Teneriffe – California – Japan – Singapore
1875–77

THE WINTER AFTER MY RETURN from Brazil I devoted to learning to etch on copper, Mr. Edwin Edwardes, who had illustrated the old inns of England, kindly giving me a few lessons. Friends seemed always accumulating round me and making life very enjoyable. The winter was an unusually cold one. After the experiences of the last two in Jamaica and Brazil I found it quite unbearable, so at last I determined to follow the sun to Teneriffe. M.E. and I started on New Year's day, 1875, in hard frost and snow, steaming from Liverpool in a wretched little steamer in unpleasant squally weather.

On the 11th we landed for a few hours in sunny Madeira. I had a cousin there with a sick husband, and in spite of the marvellous beauty of all the surroundings I pitied her for having such a number of hopeless invalids all round her. At sunset that same evening we saw the top of the Peak on the golden horizon, and on the morning of the 13th we landed at Santa Cruz.

We drove on the same day to Villa de Orotava, creeping slowly up the long zigzags leading to Laguna, where every one (who is anybody) goes to spend the hot summer months; in the New Year's time it was quite deserted, and looked as if every other house was a defunct convent. After passing Laguna, we came on a richer country, and soon to the famous view of the Peak, described so exquisitely by Humboldt; but, alas, the palms and other trees had been cleared away to make room for the ugly terraces of cacti, grown for the cochineal insect to feed on, and which did not like the shade of other trees. Some of the terraces were

apparently yielding crops of white paper bunbags. On investigating I found they were white rags, which had been first spread over the trays of cochineal eggs, when the newly-hatched insect had crawled out and adhered to them; they are pinned over the cactus leaves by means of the spines of another sort of cactus grown for the purpose. After a few days of sunshine the little insect gets hungry and fixes itself on the fleshy leaf; then the rags are pulled off, washed, and put over another set of trays. The real cochineal cactus has had its spines so constantly pulled off by angry natives who object to having their clothes torn, that it sees no use in growing them any longer, and has hardly any. These cactus crops had done another injury to the island besides that of causing it to lose its native trees. The lazy cultivators when replanting it, left the old plants to rot on the walls instead of burning them, thereby causing fever to range in places where fever had never been before; they were now planting eucalyptus-trees with a notion of driving it out.

We found there was a hotel (and not a very good one either, in its own Spanish fashion), and we got possession of its huge ball-room, which was full of crockery and looking-glasses, and some hundred chairs all piled up on the top of one another. This room had glass doors, besides other rooms opening into it, but served to sleep in well enough; and I determined to stay and make the best of it, for the climate and views were quite perfect. I did stay more than a month. Dr. Hooker had given me a letter to the Swiss manager of the Botanic Gardens, who also kept a grocer's shop. He was

very kind in taking me to see all the most lovely gardens. The famous Dragon Tree, which Humboldt said was 4000 years old, had tumbled into a mere dust-heap, nothing but a few bits of bark remaining; but it had some very fine successors about the island, and some of them had curious air roots hanging from the upper branches near the trunk, which spread themselves gradually round the surface, till they recoated the poor tree, which had been continually bled to procure the dye called Dragon's Blood. When the good people found my hobby for painting strange plants, they sent me all kinds of beautiful specimens.

After M.E. left, the landlady gave me a smaller room opening into the big room with a good view into the street, where I could live in peace and quiet, without fear of interruption, and they fed me there very kindly too. My friend the gardener arranged with the farmer at the Barenca da Castro to take me in for three days. My quarters at the old house above were very primitive. A great barn-like room was given up to me, with heaps of potatoes and corn swept up into the corners of it. From the unglazed window I had a magnificent view of the Peak, which I could paint at my leisure at sunrise without disturbing any one. I returned by a lower road, close to the edge of the sea, under cliffs covered with sedums, cinerarias, and other plants peculiar to the Canary Islands.

I stopped a while at the Rambla da Castra, on the seashore, standing almost in the sea, surrounded by palms, bamboos, and great *Caladium esculentum*. It was a lovely spot, but too glaring. After this little excursion I remained quietly working in or about Orotava till the 17th of February, when I moved down to Mr. S.'s comfortable home at Puerto di Orotava. I had a room on the roof with a separate staircase down to the lovely garden, and learned to know every plant in that exquisite collection. There were myrtle-trees ten or twelve feet high, bougainvilleas running up cypress-trees, great white lancifolium lilies (or something like them), growing high as myself. The ground was white with fallen orange and lemon petals; and the huge white cherokee roses covered a great arbour and tool-house with their magnificent flowers. I never smelt roses so sweet as those in that garden. From the garden I could stroll up some wild hills of lava, where Mr. S. had allowed the natural vegetation of the island to have all its own way. Magnificent aloes, cactus, euphorbias, arums, cinerarias, sedums, heaths, and other peculiar plants were to be seen in their fullest beauty. Eucalyptus-trees had been planted on the top, and were doing well, with their bark hanging in rags and tatters about them. I scarcely ever went out without finding some new wonder to paint, lived a life of the most perfect peace and happiness, and got strength every day with my kind friends.

Santa Cruz, to which I at first took a dislike, I found full of beauty. Its gardens were lovely, and its merchants most hospitable. I stayed there till the *Ethiopia* picked me up, on the 29th of April, with my friend Major Lanyon on board returning from the Gold Coast, where he had been filling the place of Colonial Secretary.

I got home on the 8th of May, and was soon in the full enjoyment of a London season among good friends, exhibitions, and concerts. On the 17th of July I went down to the most agreeable country house I know – that of Mr. Higford Burr, at Aldermaston. Some people I had never met before, Mr. and Mrs. S., asked me where I was going next, and I said vaguely, "Japan." They said, "You had better start with us, for we are going there also, on the 5th of August"; and, to their surprise, I said I would.

So on the 4th of August 1875 I went down to stay the night at Leasom Castle with Sir Edward and Lady Cust. The next day I went on board the *Sarmatian* at Liverpool, and found the S.s in the next cabin to myself; and he very kindly handed me in a cup of tea every morning when he made his own; for they carried every possible luxury, including canteen and box of books, and had made more journeys in less hours than any people living. We passed one or two hundred icebergs; some of them were said to be as big as the rock of Gibraltar; some of the smaller ones came too near to be agreeable. Fogs delayed us at the mouth of the St. Lawrence; we ran aground and stayed there all night till the returning tide set us free, and brought us safely up to the shore opposite Quebec. We had a cold troublesome journey through the customhouse, where my travelling companions'

Aloes (*Aloe barbadensis*) and Cochineal Cactus (*Opuntia cochinellifera*) in flower, at Teneriffe.

Above: Californian flowers featuring the crimson Snow Plant (*Sarcodes sanguinea*) from the Big Tree (*Sequoiadendron giganteum*) groves. Above, on the right, is the shrubby *Chamaebatia foliolosa* and below, on the left, the slender flower stem of *Lithophragma heterophylla*, with tufts of Spring Beauty (*Montia perfoliata*) on the right hand at the top. Two seedlings of the Big Tree can be seen behind.

Opposite: A Dragon Tree (*Dracaena draco*) at Orotava, Teneriffe. This specimen was the largest remaining after the death of the Great Dragon Tree described by Humboldt.

luggage gave them considerable occupation; the officers as usual did not even condescend to open mine. "You're a-going to paint pictures of Japan, are you? Wall! I wish you success; I should like to be going along too," the head-man said. We travelled by train to Chicago, across the prairies to Salt Lake and onwards by stage – a horrible springless machine. I had fourteen hours of it, combined with dust an inch thick all over everything. The next morning I got an old miner "guard" and a horse, left Clarks at six for the "Big Trees" of the Mariposa Grove, and had a long day's work among them.

The whole road was beautiful, through the biggest trees of the fir kind I ever saw, till I saw "The Trees." All the world now knows their dimensions, so I need not repeat them; but only those who have seen them know their rich red plush bark and the light green eclipse of feathery foliage above, and the giant trunks which swell enormously at the base, having no branches up to a third of their whole height. The little trees with wide base and tops made by shaving and narrowing the stem, which are to be found in every child's Noah's Ark, are exact models of the sequoia proportions. There were about seven hundred in that one grove of Mariposa alone, and three other groves within a day or two of them. They stood out grandly against the other trees, which in themselves would be worth a journey to see – sugar-pines, yellow-pines, and arbor vitae, hung with golden lichen. The forest was full of strange trails of big bears and other wild animals.

The descent into the Yosemite gave perhaps the very best general view of the valley; so I got our driver, after he had rested his horses and dined, to give me a lift up the hill again as far as that view, and leave me to paint it. He told Colonel and Mrs. M., who were going on with him, that "I was one of the right sort. I neither cared for bears nor yet for Ingins," and he absolutely refused to take a dollar from me when I offered it. But I had only two or three hours before dark. I could do nothing satisfactorily. The next day my friends were too tired to go beyond the verandah of the hotel; so Marie and I mounted two very "sorry nags" and accompanied a large party of tourists all round the valley to the Mirror Lake (which might have been a bit of the Tyrol), then up ladders to "Snows" a kind of "Bel Alp" hotel, which must be quite divine

A Vision of Eden

in spring from the quantity of flowers and clear water. It was a hard day's work, and the S.s "did" the Yosemite far more comfortably, and perhaps as profitably, and decided they had had enough of it, and would go back to Clarks the next day.

The same driver drove us, a most villainous-looking bandit; but he was a real good fellow, and had taken a liking for me because "I cared for neither bears nor Ingins," and he gave me some rattlesnakes' tails and a great lump of bark from the big trees, looking like a brick of solid plush. His carriage broke down with the weight of Mrs. S.'s luggage (mere necessaries! the rest having gone on to 'Frisco and £20 to pay for extra). How the driver swore (and swearing was not of a mild sort in California), then he turned round quite gently to me: "Now don't you go for to take any of them lazy cattle guides to 'The Trees' again;

The "Great Grisly" Big Tree (*Sequoiadendron giganteum*) in the Mariposa Grove, California, U.S.A.

you are going a long journey, and it's the dollars you want; don't you waste them on such brutes. I'll tell Moore to give you a good old 'orse as I knows the ways of, and show you how to loose his girths, and you just stay and draw till you're tired, and tie 'im up and loosen 'im, and then tighten 'im again, and come 'ome quiet; and if you don't say nothing to nobody, nobody won't say nothing to you; you'll save your dollars, and that's what you want." So I did say nothing to nobody, because I never saw anybody to say anything to all day after the S.s went. I had a long day's work in that lovely forest painting the huge tree called the Great Grisly, whose first side branch is as big as any trunk in Europe. After that I went down to 'Frisco and became No. 794 in the Occidental Hotel.

The hotel was admirably managed, with lifts to every storey, as well as grand staircases. In the afternoon the consul called for me with Colonel and Mrs. M. (whom I had just seen in the Yosemite), and he drove us on the top of a pair of spidery wheels to Cliff House, to see the Islands of Sea-lions, or seals. Those rocky islands were some hundred yards from the balcony of the hotel, which had been built for the purpose of feeding and sheltering the cockneys of 'Frisco, who often spend a "happy day" in watching the crowds of sea-beasts through various telescopes which are fixed for the purpose. The American Government protects these creatures, and no boat is allowed to go near them, or any shooting practised.

The next day I returned and spent the day painting at Cliff House, and the day after that I started back to the "Summit Station," Colonel and Mrs. M. going with me as far as Sacramento, where there was a fair at which he hoped to see fine horses and cattle, but was disappointed. I continued in the train, which slowly climbed its 8000 feet and landed me at midnight at the top of the pass, in the midst of the Nevada Mountains, and I settled for a week in a very comfortable railway-hotel. Half-an-hour's climb took me to the highest point near, from which was a most magnificent view of the Donner Lake below, and all its surroundings. Of this I made two large sketches, taking out my luncheon, and spending the whole day on those wild beautiful hills, among the twisted old arbor vitae, larch, and pine trees, with the little chipmunks (squirrels) for company, often not bigger than large mice.

An old cypress or juniper tree in the Nevada Mountains, California, U.S.A.

My landlord drove a drag, four-in-hand, down to Lake Tahoo most days, and at the end of the week took me on there, driving down the steep descent to Lake Donner. We went along the whole length of its clear shore to Truckee, then followed the lovely clear river to its source in the great Lake Tahoo, a most lovely spot with noble forests fringing its sides. There was another capital wooden hotel there, where I could work again in peace. Behind the house were noble trees, fast yielding to the woodman's axe; huge logs were being dragged by enormous teams of oxen, all smothered in clouds of dust. They made fine foregrounds for the noble yellow pines and cypress-trees, with their golden lichen. The M.s picked me up there again, and after going round the lake in the little steamer we disembarked on the east side, and took a carriage with a driver who has been made famous by Mark Twain. Two

hours' rail at the end of this drive took us up the hills to Virginia City. From there I went back to the Summit Hotel, which I reached at four in the morning. After a few days I descended to Stockton, and by another line to Milton, thence by stage to Murphy, which I did not reach till nine at night in the dark.

The next morning I drove on to Calaveras Grove, found myself the last guest of the season in the comfortable hotel under the big trees, and stayed there a week. That was indeed luxury, to be able to stroll under them at sunrise and sunset without any delay or trouble. A stag with great branching horns was my only companion; he had a bell round his neck, and used generally to live in front of the house, but liked human company; and when I appeared with my painting things he would get up and conduct me gravely to my point and see me well settled at work, then scamper off,

87

coming back every now and then to sniff at my colours. One of my first subjects was the great ghost of a tree which had had a third of its bark stripped off and set up in the Crystal Palace; the scaffolds were still hanging to its bleached sides, and it looked very odd between the living trunks of red plush on either side. The sugar-pines were almost as large, and even more beautiful than the sequoias, their cones often a foot long, and so heavy that they weighed down the ends of the branches, making the trees look like Chinese pagodas in shape. They are called sugar-pines from the white sweet gum which exudes from the bark, and drops on the ground like lumps of brown sugar; it is much eaten by the Indians. The cones of the "Big Trees" were small in proportion.

My time was up, and I had to go back to 'Frisco and civilised life. I was very anxious to see some of the red-wood forests. They had been so destroyed that it was not easy to get to them, but the village doctor gave me a letter of introduction to the head woodman of an estate, some two or three hours up the northern line of rail, who took me to his house to sleep. It was only a small hut of logs, but they had a spare room, and made me very welcome. The red-wood trees are all about those hills, and are more like silver-fir than the other sequoias. My host took me some miles up a side valley to see some which were fifteen feet in diameter, and nearly 300 feet high. They were gradually sawing them up for firewood, and the tree would soon be extinct. Its timber is so hard that it sinks in water, and no worm can eat it there. It is invaluable for many purposes, and it broke one's heart to think of man, the civiliser, wasting treasures in a few years to which savages and animals had done no harm for centuries. I settled myself to sketch near a "bear's bath," hoping to see the big beast come and wash himself, but he didn't. I saw two pretty little deer and numbers of squirrels and birds, then walked back and was put on the engine of a wood train, as the passenger train had gone by some hours before.

On the 16th of October I took possession of a splendid, large, airy cabin in the *Oceanic*, one of the finest steamers afloat, fitted up in the most luxurious way, with an open fireplace in a corner of the great saloon, which we were very glad of after the first week, as we went by the northern route, which was too cool for pleasure. Three

Distant view of Mount Fujiyama, Japan, and the beautiful climbing shrub *Wisteria chinensis* – a native of China and Japan.

weeks without seeing land at all is a long time, and latterly I suffered much from an attack of my old pain, brought on by the cold. We jumped in one day from the 28th to the 30th of October, and at daylight on the 7th of November found ourselves within sight of Fujiyama. I watched the sun rise out of the sea and redden its top, as I have seen so well represented on so many hand-screens and tea-trays. Some of my ship friends landed with me. We drove out into the country, and took funny cups of yellow tea in a bamboo tea-house, with five pretty girls rather over four feet high, in chignons with huge pins, blackened teeth, and no eyelashes, laughing at us all the while.

The next morning I saw my friends off. The big ship departed, and I returned to the hotel at Yokohama – a sort of mongrel establishment, with neither the cleanliness of Japanese nor the comforts of English life. Mrs. C. soon found me out. As Sir Harry and Lady Parkes were said to be

soon going away on an expedition round the coast, I started to pay my respects to them at eight in the morning. The railway went alongside the famous Tokada Road, and was full of interest. The rice and millet harvest was then going on, and the tiny sheaves were a sight to see. They piled them up against the trees and fences in the most neat and clever way, some of the small fan-leaved palm-trees looking as if they had straw petticoats on. There was much variety in the foliage; many of the trees were turning the richest colours, deep purple maples and lemon-coloured maidenhair trees (*Salisburia*), with trunks a yard in diameter. The small kind of Virginian creeper (*Ampelopsis*) was running up all the trees. These seemed generally dwarfed, except round the temples, which were marked all over the country by fine groves of camphor, cryptomeria, cedars, and pine-trees, as well as a small variety of bamboo.

At the last station one of the Japanese ministers got into our carriage in the costume of a perfect English gentleman, chimney-pot hat included. He invited me to come and see his wife at his country-house, and at Yedo packed Miss C. and myself into two jinrickshas, a kind of grown-up perambulator, the outside painted all over with marvellous histories and dragons (like scenes out of the Revelation). They had men to drag them with all sorts of devices stamped on their backs, and long hanging sleeves. So we trotted off to the Tombs of the Shoguns, most picturesque temples, highly coloured and gilded, half buried in noble trees, under a long low ridge or cliff. We left our cabs, and wandered about amongst them attended by a priest, a wretched mortal who would have sold even Buddha himself for a few cents if he dared run the risk of being found out.

After a few days' quiet at the British Legation, I started in the steamer for Kobé, another of the European settlements of Japan – a pretty place on a quiet bay of the sea, with high hills behind it, and an interesting temple, at the entrance of which was a shed with a white horse in it of a peculiar breed, with blue eyes and pink nose, and hoofs turned up from want of exercise. This horse was kept in case God came down and wanted a ride. Plates of beans are put on a table near, with which pious people feed the horse as they pass in, dropping some

money into a box at the same time to pay for them. A stuffed horse is kept in another shed close by, to be ready, in case the holy beast should die, to fill his place, and not disappoint the equestrian Deity.

Kobé was a very sociable place. Lady Parkes was not sorry to make me an excuse for escaping its heavy luncheons and dinners, and we started by rail for Osaka, where we took jinrickshas, with a tandem running in front, and trotted about ten miles to the valley of Minbo, famous for its maples. Lady Parkes and her two A.D.C.s went back to endure a state dinner at Kobé, while I made my way to the inn (kept by a Frenchman), appointing to meet Sir Harry and his party at the railway at nine the next morning. We started with our luggage in fifteen jinrickshas, with two men in each, one in the shafts and one running tandem in front. They trotted over thirty miles that day. They never got in the least tired, but did the last part of the way up the High Street of Kioto at a gallop after nearly seven hours of hard running. The road was generally very narrow; the bridges, placed at right angles to it, rather steep up and down and without parapet, were very disturbing to one's nerves, as the men never broke their pace, but swung one on and across and round again in one even jog. We passed through the richest cultivation – rice, tea, buckwheat, cotton, mulberries, bamboos, camellias twenty feet high, full of single pink and white blossoms. Oranges, persimmons, and Japan medlars seemed the common fruits.

Before dusk we were in the long suburbs of the old capital of the Tycoon. Our men went faster and faster, till they nearly galloped us up the long High Street and steep ascent to the hotel; and soon after the Governor of Kioto (in corduroys and shooting-jacket, and about four and a half feet high) appeared to pay his respects to Sir Harry and beg us all to go and dine with him. His Excellency

begged to be excused, but promised to have luncheon with him the next day at a tea-house the other side of the valley, for the Governor was starting on an official journey, and that would be on his way. We worked hard next day, and saw many wonderful temples and palaces all of wood, with a beautiful concave curve in their overhanging roofs.

We drove out to the Governor's luncheon party at the tea-house, which had one side of the room quite open towards a pretty garden and a clear view. On the table were vases of chrysanthemums, tied on all the way up sticks a yard high, so as to show all the flowers and hide the stalks. The ornaments were of rare old Satsuma porcelain; the food which came from our hotel, being of the knife-and-fork order, not interesting. After luncheon we took leave of the Governor and pulled up the river, getting out where the valley narrowed to walk along its banks. I saw the leaves of *Primula sinensis* and ferns, but there were few flowers at that season. We also saw many lovely kingfishers.

The next day Sir Harry and Lady Parkes and their suite departed, leaving me in sole possession, with a special order from the Mikado to sketch for three months as much as I liked in Kioto, provided I did not scribble on the public monuments or try to convert the people; for it was still a closed place to Europeans. Sir Harry himself had been nearly murdered on his last visit there, and Sir Rutherford Alcock was never even allowed to enter. But I was perfectly safe all alone, and comfortable too, in the old temple building some centuries old, which had been turned into an hotel for Europeans, with the addition of a few chairs and tables. My room was made of paper, with sliding-panels all round, two sides opening to the frosty air and balcony, the other two only going up about seven feet, leaving abundant ventilation between them and the one great roof of the whole house, with the advantage of hearing all my neighbours' conversation beyond. From my windows, when I pushed back the paper sliding shutter, I saw a most exquisite view (for the house was perched up high on the side of the hill, with the most lovely groves and temples all over it), and below the great city of over 200,000 inhabitants. The top of one of the favourite trained pinetrees came up like a terrace of flat turf to the level of the balcony; it looked so

Above: The Hottomi Temple, Kioto, Japan, with a pine tree trained as an espalier in front and autumn tints in the background.

Below: Japanese flowers. A yellow-flowered *Forsythia suspensa*, with members of the genera *Rhododendron* and *Camellia* plus *Paeonia suffruticosa*. The variegated foliage belongs to *Cleyera fortunei*.

solid that I could almost have walked over it. Groups of gray boulders, and small clipped azaleas, heaths, and camellias, with many other flowers and small tufts of pampas-grass and bamboo, filled the rest of the garden, varied by little miniature lakes and canals. The drawback to me was the cold, which was intense at night. The charcoal pans (most classically shaped) were a poor substitute for fires, but the ventilation and draughts of the rooms were so great that one was in no risk of suffocation from the fumes. After sketching all day amongst the dead leaves, and morning white frosts, I used to be scarcely able to stand from stiffness and coming rheumatism, and had to hold on by a tree at first, till I could use my feet.

I went to a place where they sell live pets, and saw the most beautiful gold and silver pheasants, mandarin ducks, monkeys, and gazelles, and hideous brown salamanders from Lake Biwa, two feet long; also tortoises in a tank. The tortoise with a green tail Japanese are so fond of embroidering, is merely one with green algae growing on its shell. Japan was most attractive. There was always something new and interesting to meet me every day. I had hoped to stay over the winter, and to go to the hills and Nikko in the summer, but I got stiffer and stiffer, and at last could scarcely crawl; so on the 19th of December I ordered a boat to Osaka, and set myself to pack as well as I could, with a fool to help me, crippled hands, and bones full of pain. From Osaka I went by train to Kobé, then I took the steamer the next day for Yokohama, where the C.s again received me. I was in the doctor's hands for ten days with rheumatic fever. I could not even feed myself during part of the time.

Mrs C. was extremely kind, went on board the steamer with me, and secured me a good cabin to myself. The Messagerie boats are certainly the very best in the world. That one was so beautifully warmed and sweet, that it seemed like a change of climate when I entered it. I got better every day, all was so clean and the cooking so good. We had beautiful calm weather, and entered the harbour of Hong Kong about eight in the evening, when its semicircle of lights were bright as the stars above. We all moved our things out of the nice little ship we had come in to a larger one of the same company on the other side of the harbour,

and passed on our way the steam-launch Commodore Parish had sent for me; but it soon followed and took me on board the old Hospital Ship, the *Victor Emmanuel*. The Commodore took me on shore. We did not do more than we could help, but saw enough to give me an idea of how pretty those hills might become in a few years by irrigation and good management. If I had only been well I could have stayed on there, and gone the next week in the Commodore's steam-launch up to Canton, one of the wonders of the world; but it was wiser to get nearer the equator, and four days more took me into heat enough at the French settlement of Saigon, and the mouths of the river which leads to that wonderful old forest full of ruined palaces and temples, Cambodia, about which so little is known.

Two more days brought us to Singapore, where I landed on the 19th of January 1876. I could barely hobble from the office of the hotel to my rooms at the other end of the building, through its lovely garden; but how delicious that still warm air was, with exquisite blue sky, lilac shadows, and white lights! I found a lemon-tree close to my room, covered with tailor-ants which had sewn up the leaves into most ingenious nests, the pretty flowers opening their sweet petals close to them.

One of my windows was quite blocked up by a great bread-fruit tree covered with fruit as big as melons, with leaves two feet in length, gloriously glossy, and I set myself at once to make a study of it. While at this work Mrs. S., the banker's wife, and her father, Major MacN., came to see me. The former insisted on my moving at once to her comfortable house outside the town. Like all the houses of Singapore, it stood on its own little hill, none of these hills being more than two hundred feet above the sea; but they were just high enough to catch the sea-breezes at night, and one could sleep with perfect comfort, though only three degrees from the equator. The lovely Mangosteen was just becoming ripe, and the great Durian, which I soon learnt to like, under the teaching of the pretty little English children, who called it "Darling Durian." No garden could have delighted me more than that one. Every day I was sure to

Bread-fruit (*Artocarpus altilis*) and butterflies (*Papilio polytes* and a white *Pieris*) painted in Singapore.

A Vision of Eden

find some new fruit or gorgeous flower to paint, Mrs. S. working beside me all the hot day through in her deliciously airy upper rooms.

The Botanical Garden at Singapore was beautiful. Behind it was a jungle of real untouched forest, which added much to its charm. In the jungle I found real pitcher-plants (*Nepenthes*) winding themselves amongst the tropical bracken. It was the first time I had seen them growing wild, and I screamed with delight. One day we drove out to have luncheon with the Doctor and his family, who had a country house about five miles off, near the coast, in the midst of plantations of cocoa-nuts. One wild plant I saw there for the first time, the *Wormia excelsa*, which abounds in the different islands of "Malaysia," and is often planted as a hedge. It has a glossy five-petalled flower of the brightest yellow, and as large as a single camellia, with large leaves like those of the chestnut, also glossy, and separate seed-carpels which, when the scarlet seeds are ripe, open wide and afford a most gorgeous contrast of colour with its waxy green and scarlet buds. I know few handsomer plants. All the tribe of Jamboa fruits (magnified myrtles), too, were magnificent in their colours. There were said to be three hundred varieties of them.

Mrs. S.'s verandah was full of rare plants; orchids, caladiums, and other exquisite things. On one, *Ficus benjamina*, were planted some score of phaloenopsis in full flower, like strings of white butterflies hovering in the air with every breath of wind.

After a fortnight I went to stay at Government House with Sir William and Lady Jervois. Close under my window was a great india-rubber tree with large shiny leaves and fantastic hanging roots. In the front of the garden was a gorgeous tree of *Poinciana regia* blazing with scarlet blooms. I immediately begged a branch and hung it up to paint, but made a most absurd mistake. I placed it the wrong way up. It was stupid, but I was consoled afterwards when I found that that clever Dutch lady, Madame van Nooten, had actually published a painting of the poinciana growing in the same topsy-turvy way! Nothing approaches this tree for gorgeousness; the peculiar tender green of the acacia-like leaves enhances the brilliancy of its vermilion tints. It was curious to see how little the English people cared for these glories around them. Lawn-tennis and croquet were reigning supreme in Singapore, and little else was thought of after business was over.

94

CHAPTER VI

Borneo and Java

1876

AFTER A FORTNIGHT AT GOVERNMENT HOUSE, Sir William wrote me letters to the Rajah and Rani of Sarawak, and I went on board the little steamer which goes there every week from Singapore. After a couple of pleasant days with good old Captain Kirk, we steamed up the broad river to Kuching, the capital, for some four hours through low country, with nipa, areca, and cocoa-nut palms, as well as mangroves and other swampy plants bordering the water's edge. On the right bank a flight of steps led up to the Rajah's Palace. I sent in my letter, and the Secretary soon came on board and fetched me on shore, where I was most kindly welcomed by the Rani, a very handsome English lady, and put in a most luxurious room, from which I could escape by a back staircase into the lovely garden whenever I felt in the humour or wanted flowers.

The Rajah, who had gone up one of the rivers in his gun-boat yacht, did not come back for ten days, and his wife was not sorry to have the rare chance of a countrywoman to talk to. The Rajah was a shy quiet man, with much determination of character. He was entirely respected by all sorts of people, and his word (when it did come) was law, always just and well chosen. He had one hundred soldiers, a band which played every night when we dined (on the other side of the river), and about twenty young men from Cornwall and Devonshire called "The Officers," who bore different grand titles, – H. Highness, Treasurer, Postmaster-General, etc., – and who used to come up every Tuesday to play at croquet before the house.

The views from the verandah and lovely gardens, of the broad river, distant isolated mountains, and glorious vegetation, quite dazzled me with their magnificence. What was I to paint first? Every one collected for me as usual. Orchids and pitcher-plants were pulled for me most ruthlessly, the latter being of several varieties, from the tiny little plants which grew in the meadow near, and whose pitchers were not half the size of thimbles, to trailing plants of six or eight feet long. The common pepper-plant, too, was much cultivated and very elegant, as well as gambier and other dyes, sago, and gutta-percha, the former growing thirty feet high, with grand terminal bunches of flowers from the centre of the crown (very unlike the small cycads people had called the sago palm in other countries). It takes fifteen years before it flowers; then, before the fruit has time to ripen, the whole tree is cut down and the pith taken out and washed. Wallace says one tree could supply a man with food for a whole year. The gutta-percha trees were fast disappearing. They ought to have been protected by law, and the people compelled to bleed them as in other countries, not to sacrifice the great trees for one crop – trees which had been a hundred years growing, and could not be quickly replaced.

There were acres of pine-apples, many of them having the most exquisite pink and salmon tints, and deep blue flowers. The mangosteen, custard-apple, and granadilla were also in abundance. The mangosteen was one of the curious trees people told me never had a flower. But I watched and hunted day by day till I found one, afterwards seeing whole trees full of blossoms, with rich

A Vision of Eden

Above: An old boathouse and riverside vegetation in Sarawak.

Opposite: Nipa or Nypa Palm (*Nypa fruticans*) in Sarawak, with an inflorescence in the foreground, part of a leaf behind and a mature tree in fruit in the distance.

crimson bracts and yellow petals, quite as pretty as the lovely fruit. This last is purple, and about the size of an orange, with a pink skin inside, divided into segments, six or more, which look like lumps of snow, melting in the mouth with a grape-like sweetness. The duca was a still finer fruit of the same order, growing in bunches, with an outer skin or shell like wash-leather, and a peculiar nutty flavour in addition to its juiciness. The custard-apple was well named, for it is a union of both words. Its outside is embossed with lozenges of dark green on an almost creamy ground, and over the whole a plum-like bloom, very difficult to paint, and indescribably beautiful.

There was a magnificent specimen of the Madagascar ravenala or travellers' tree, close to the house on the other side of a small bend of the river, and the Rajah had had the good taste to leave all its younger off-sets round it uncut. I spent some afternoons in drawing that view, and used to see numbers of graceful water-snakes swim up the

View of the hill of Tegoro, Sarawak.

giant trees as foreground round the clearing, which was also full of stumps and fallen trees grown over with parasites – the most exquisite velvety and metallic leaves, creeping plants, "foliage plants," caladiums, alpinias, and the lovely *Cissus discolor* of all manner of colours, creeping over everything.

Great parasitic trees were standing there with their stalks all plaited together. They had strangled the original tree on which they had lived the first years of their treacherous lives, and were now left like tall chimneys of lattice-work, their victim having rotted away from the centre. There were masses of tree-ferns; one group round a little trickling spring which supplied the house with water, I could not help painting. Life was very delicious up there. I stayed till I had eaten all the chickens, and the last remains of my bread had turned blue; then, having seen the smoke of the parting salutes through my telescope in the swamp far below, I came down again, my soldier using his fine long sword to decapitate the leeches which stuck to me by the way. I had a most enjoyable day; for we hunted up all sorts of orchids, pulling under the thick overhanging trees, while the boy ran up the branches like a monkey, cut them through with the soldier's silver-mounted sword, and let the tangled masses tumble down into the water below with a great splash and a flop, nearly swamping my small nut-shell of a canoe. We picked off all the treasures, and soon had a perfect haycock of greenery in the middle of the boat, to carry home and hang up to the Rajah's trees in the garden.

The Rajah was very glad of all the things I brought; but hanging on a dry tree over a well-mown grass lawn is a very different thing from living in a swamp over the water, and I fear few of my treasures lived long. He and the Rani went one expedition with me in the yacht, first going down the river to the sea, then up in a boat over the sand-bar, and up another big river, past groves of casuarina-trees, winding in and out, almost back to the sea again, as far as Loon Doon, where we found a very nice house, and the magistrate, Mr. N., a most hospitable host. The forests behind his house were really magnificent. *Clerodendron fallax*, whose blooms used to be employed by the Dyaks to dress the heads of their enemies taken in battle, and the large kinds of mussaenda, were

creek with their heads curved well out of the water. Iguanas I also saw, and monkeys which used to come down to the edge of the garden and laugh at us. Sweet singing-birds were very plentiful.

One day a letter came, announcing that Captain Buller, R.N., was going to bring the new Consul of Labuan in his war-ship to pay a visit to the Rajah; so, as his spare rooms were only two, I persuaded him to send me off out of the way to his mountain-farm at Mattange. The Rajah lent me a cook, a soldier, and a boy, gave me a lot of bread, a coopful of chickens, and packed us all into a canoe, in which we pulled through small canals and forest nearly all day; then landed at a village, and walked up 700 feet of beautiful zigzag road, to the clearing in the forest where the farm and châlet were. The view was wonderful from it, with the great swamp stretched out beneath like a ruffled blue sea, the real sea with its islands beyond, and tall

The quicksilver mountain of Tegoro, Sarawak, by moonlight.

particularly striking, the white bracts of the latter catching one's eyes at every turn.

The Rajah had planned taking me to some other stations, but his wife was suffering. We went back instead, and soon after I started in the small steam-launch up the river, with Mr. B., the good-natured Scotch-manager of the Borneo Company's mines in Sarawak, and a young Devonshire giant, rejoicing in the title of "His Highness the Rajah's Honourable Treasurer." The banks of the river were a continual wonder all the way up, with creeping palms or rattans binding all the rest of the greenery together with their long wiry arms and fish-hook spines. I traced this plant far up into the high trees. No growing thing is more graceful or more spiteful. We slept at some antimony-mines near the river. I found that the manager and his wife came from Hastings, and their belongings lived in its old High Street, and knew myself and my dear old father perfectly well by sight. We

continued our journey next morning on a springless tram-cart, with our feet hanging down behind, and considered that a rest for three hours; after which an excellent little pony carried me, while the men walked, through a marvellous forest for fifteen miles, except when we came to the broken bridges and I had to balance myself on cranky poles while the pony scrambled through below. Some of the way was under limestone cliffs.

The forest was a perfect world of wonders. The lycopodiums were in great beauty there, par-ticularly those tinted with metallic blue or copper colour; and there were great metallic arums with leaves two feet long, graceful trees over the streams with scarlet bark all hanging in tatters, and such huge black apes! One of these watched and followed us a long while, seeming to be as curious about us as we were about him. When we stopped he stopped, staring with all his might at us from

99

behind some branch or tree-trunk; but I had the best of that game, for I possessed an opera-glass and he didn't, so could not probably realise the whole of our white ugliness.

I never saw anything finer than the afterglow at Tegoro. The great trees used to stand out like flaming corallines against the crimson hills. It was lovely in the full moon, too, with the clouds wreathing themselves in and out of the same giant trees around us. We had our morning tea at half-past six on the verandah, and a plum-pudding in a tin case from Fortnum and Mason was always brought out for the benefit of our young Cornishman, who was always ready for it.

Mr. B. walked him off after it, and I had all the day in perfect quiet to work in the wild forest or the verandah on different curious plants. One creeper with pink waxy berries like bunches of grapes was particularly lovely; and the scarlet velvet sterculia seed-case, with its grape-like berries, most magnificent in colour. Mr. B. soon started for some other mines, and I was left to the care of his assistant, Mr. E., who had been sent out originally as a naturalist by Sir Charles Lyell, in search of the "missing link," or men with tails; and after searching the caves in vain, kept himself alive by "collecting" for different people at home. Mr. B. found him out and sent him to Tegoro. He was full of wit and information about the country. I found him a most delightful companion, as good as a book to talk to, and he was delighted to find one who was interested in his hobbies.

At last I had to leave Tegoro. Mr. E. walked down two miles with me; then we got into a canoe and shot the rapids for many more miles, with the great trees arching over the small river we followed, and wonderful parasites, including the scarlet aeschynanthus, hanging from the branches in all the impossible places to stop. We sat on the floor of the canoe, held on tightly, and went at a terrible pace, the men cleverly guiding us with their paddles and sticks. At last we got out and walked again through the wonderful limestone forest and out to the common or clearing round Jambusam, where there was a long-forsaken antimony-mine. Mr. B. had kindly arranged for me to stay there, and had sent food and furniture to meet me.

Mr. E. went up a mountain near and brought me down some grand trailing specimens of the largest of all pitcher-plants, which I festooned round the balcony by its yards of trailing stems. I painted a portrait of the largest, and my picture afterwards induced Mr. Veitch to send a traveller to seek the seeds, from which he raised plants and Sir Joseph Hooker named the species *Nepenthes northiana*. These pitchers are often over a foot long, and richly covered with crimson blotches. Then I said good-bye to Mr. E. and returned to the Rajah's at Kuching.

The next day I said good-bye to His Highness, who came on board to see me off at seven in the morning, like the real English gentleman he is. Mr. B. also went across to Singapore with old Captain Kirk, and we were a pleasant little party of three on deck. The weather was so calm and warm that we had our meals under the awning. Those two would not let me land in the ordinary way, but made me wait till the Company's boat fetched me, with a grand native in a gorgeous turban to look after my luggage, and put me into somebody's smart open carriage, which conveyed me with great dignity to Government House.

Lady Jervois had sent to meet me by the last mail, and this one was before its time; but she made me very welcome, and I stayed there till the Java steamer started – a most comfortable Messagerie boat with few passengers, but a most entertaining monkey belonging to the captain. There was one Englishman only on board. He remarked that he thought it was very hard that the little beast should have the luxury of re-enjoying his dinner whenever he chose to take it out of that great cheek-pouch, thus having one pleasure more than human beings. He also contradicted me flatly when I talked of the *Amherstia nobilis* as a sacred plant of the Hindus. I said I thought Sir W. Hooker told me it was so, and he said Sir William had been a great botanist, but was not a Hindu scholar. I had made a mistake, and I began to look at the little man with respect, and found he was Dr. Burnell, the famous Indian scholar and Judge of Tanjore, making a pilgrimage to Boro-Bodo during his short spring holiday; so we became friends, and continued so till he died. I like a real contradiction when it has a reason behind it, and there were plenty of reasons in Dr. Burnell.

Opposite: Pitcher plants from Sarawak – *Nepenthes rafflesiana* with below *N. ampullaria*.

Borneo and Java

When we reached the roads off Batavia we were transferred to a small steam-launch, which took us for a couple of miles through a long walled canal with sea on each side beyond the walls. At the Custom House my friends handed me over to the care of its Head, who would not look at my luggage, but told me to wait a little till the train started for Buitenzorg. After an hour, during which time I sat still on my trunk sketching boats and banana-trees, he returned to tell me the train had gone an hour ago, and there was no other till the next day; so he packed me and my trunks into the smallest of dog-carts, with a mite of a pony to draw it, which I expected to see lifted off its legs by the weight behind. It took some time to start the poor little beast off, but being once set going, he dashed at a furious pace all the way to the hotel, which consisted of a straggling collection of ground-floor rooms, with verandahs and sleeping men on rocking-chairs all round them in the lightest possible clothing. About five o'clock I put on my best dress and took my letter to the President of the Council, M. van Rees, a most courteous and agreeable man. His wife was in the hills, where he said I must go and see her, and he handed me back to the carriage as if I were a princess, and told the driver where to take me so as to have an idea of the outside of Batavia.

Batavia was a most unpleasant place to sleep in, full of heat, smells, noise, and mosquitoes. I started as soon as possible the next morning in the train for Buitenzorg, which, though only a few hundred feet above the sea, has pure cool air at night. Every one (who is anybody) has a villa there, and merely goes to the city on business and as seldom as possible. The old French landlady said she had been expecting me a long while, and gave me a cheerful little room with a lovely garden on each side, with such cocoa-nut, breadfruit, and bananas that it was a real joy to sit still and look at them; and I resolved to stay quiet for a month or more, and learn a little Malay before I went anywhere else. Mr. and Mrs. F., who lived close by in the most exquisite little garden that ever was seen, promised

to make all easy for me both at Buitenzorg and on my future travels, and they abundantly fulfilled their promises.

The order of everything in Java is marvellous; and, in spite of the strong rule of the Dutch, the natives have a happy, independent look one does not see in India. Java is one magnificent garden of luxuriance, surpassing Brazil, Jamaica, and Sarawak all combined, with the grandest volcanoes rising out of it. These are covered with the richest forests, and have a peculiar alpine vegetation on their summits. One can ride up to the very tops, and traverse the whole island on good roads by an excellent system of posting arranged by Government. There are good rest-houses at the end of every day's journey, where you are taken in and fed at a fixed tariff of prices. Moreover, travellers are entirely safe in Java, which is no small blessing.

The sago-palms were just then in full flower, with great bunches of pinkish coral branches coming out of the centre of their crowns. The fruit when ripe is like green satin balls quilted with red silk.

The famous Botanic Garden was only a quarter of an hour's walk from the hotel, and I worked there every day, but soon found it was of no use going there after noon, as it rained regularly every day after one o'clock, coming down in sheets and torrents all in a moment. The Governor-General asked me to dinner in his grand palace in the midst of the garden. There were several people there, and some great men with fine orders on their coats; and when a little dry shy-mannered man offered me his arm to take me in to dinner, I held back, expecting to see the Governor-General go first; but he persisted in preceding the others, and I made up my mind that Dutch etiquette sent the biggest people in last, only taking in slowly that my man *was* his Excellency after all.

There was another hotel in the place, with a most magnificent view from its terrace, which I painted, looking over miles of splendid plantations of cocoa-nut and every kind of fruit-tree, with patches of rice and other grain between, leading up through grand forests to the most stately volcano, with a wide river winding underneath, full of people wading, washing, and fishing. Those amphibious people always prefer to go through the water rather than over it on bridges, and they go in, clothes and all, in the most

Wild flowers of Sarawak. In the centre *Coelogyne asperata* with the young inflorescence of a member of the ginger family behind; the pitchers belong to a *Nepenthes* species, with the purple-flowered *Dendrobium secundum* lying in front.

A Vision of Eden

decent way. Men and women dress almost alike, in all the brightest colours, with rich Indian scarves thrown round them, and always the inseparable umbrella.

The Dutch food in Java was peculiar, but good, the principal meal being at twelve o'clock, when one found on every plate a mountain of well-boiled loose rice. To that one added chicken bones, rissoles, sausages, cutlets, poached eggs, salt fish, curry sauce, stewed bananas, and a dozen other incongruous things, and ate them all together with a spoon. They also had beef-steaks and potatoes in some places, and dessert; but the former was often of buffalo flesh, which is blue and black, not tempting, and one had little desire for more food after the first mixture. The evening meal was a much lighter one. In the afternoon at four o'clock cups of tea were taken into every room, and the world dressed itself up to pay visits and walk or drive. The baths, too, in those regions are taken in an odd but very agreeable manner, in marble baths with the water coming from a spout overhead, and running out at the bottom, merely

Above: View from Buitenzorg of the Salak Volcano, Java, clothed mostly in rich forest, but with patches of cinchona, coffee and tobacco cultivation on the upper slopes, with, lower down, terraces of rice and Indian corn.

Opposite: A group of wild flowers of Java from Tosari. These flowers were from an elevation of 6,000ft. The most noteworthy flower in the group is the fine large Forget-me-not in the centre. Two of the plants in front on the left, the *Physalis* and *Datura* (long purple flower), are common in nearly all warm countries. The yellowish-flowered *Hydrangea*, in front of the young fern-fronds, indicates some relationship with the vegetation of Japan. Other genera represented are *Impatiens*, *Melastoma*? and *Crotalaria*; the small blue peaflower hanging in front is the Shamrock Pea (*Parochetus communis*).

splashing one all over, with a bit of perforated wood to stand on. It refreshes one much more than soaking in water. There were abundance of baths in the hotel, and they could be taken at any hour.

The Botanic Garden was a world of wonders. Such a variety of the different species was there? The plants had been there so long that they grew as if in their native woods – every kind of rattan, palm, pine, or arum. The latter are most curious in their habits and singular power of emitting heat. All the gorgeous water-lilies of the world were collected in a lake in front of the palace. The Director was most kind in letting me have specimens of all the grand things I wanted to paint. The palms alone, in flower and fruit, would have easily employed a lifetime. The blue thunbergia and other creepers ran to the tops of the highest trees, sending down sheets of greenery and lovely flowers.

After more than a month at Buitenzorg I left my heaviest trunk and started for Batavia, with a big letter in my pocket from the Governor-General to all officials, native and Dutch, asking them to feed and lodge me, and pass me on wherever I wished to go. I found the hotel at Batavia quite full, but the landlord and landlady made me live in their rooms and eat with them, most kindly putting me up for the night "somehow," and charging nothing.

Mrs. F. had kindly arranged with the captain of

The Soembing Volcano, Java, seen from Boro-Bodo, with a plantation of teak trees (*Tectona* sp.) and the artist's home for ten days in the foreground.

my ship to call for me at six one morning and take me with him on board. The ship was full of great people, three Residents (Lord-Lieutenants of Counties) and a Colonel of Engineers going with his men to make a railway. At Samarang every one but myself and the captain went on shore, but as I was to return that way I preferred staying quiet and painting the glorious view of its harbour and the five volcanoes from the deck. There was no snow, but they were all about 10,000 feet high, and their slope was steeper than that of Etna or Teneriffe. One smaller one was still smoking, the others were quiet.

I only stayed a night in the hotel at Soerabaja, then Mrs. F.'s nephew, the town-clerk of the place, took me off to his house in the suburbs (where everybody had their villas), and gave me a delicious room in his garden. His wife Mrs. S.H. was most hospitable. They were the first people who had really shown that virtue, though many talked of it. I promised to return straight to her house after my expedition to the mountains. I drove from their house "post" with four horses, which went full gallop and were changed every three or four miles without a moment's loss of time (as well as their coachman and groom, to whom I always gave a fee of twopence each), through an almost continuous avenue of tamarind trees, which met overhead, shading the long straight road most deliciously. This was mended and swept as smooth as a carpet, the bullock-carts and heavy traffic being forced to go on a parallel road outside the trees.

I stayed three days in an excellent hotel at Pasoerocan and from there went on to Tosari which is 6000 feet above the sea. Its season was over, and it was cold at night, and generally wrapped up in clouds. The scenery is very curious, the steep volcanic hillside ploughed up into great furrows from top to bottom, often 1000 feet deep, and the tops a few yards across. One could talk to people on the opposite hill-slope, though it would take hours of hard scramble or roundabout paths to reach them.

My landlord was a most entertaining companion, speaking perfect English, and knowing the whole place well. He and his wife were both musicians. They had a piano and harmonium, and sang really well. He also did a good deal of doctoring, giving ten grains of quinine or an electric shock from his machine in exchange for a fat chicken. This the natives considered a fair exchange.

It seemed like leaving home again to come away from those kind people. I had been told in the plain the road I wished to go was full of difficulties and dangers, but I found none. At last we got to Pakis, and rode to the house of the chief, with a letter from my Tosari landlord asking him to send me on to Malang. The chief informed me I should be sent on soon, and a good deal more in a more than unknown tongue; for in that part of Java they talked Javanese, not Malay, and the former language I had not even attempted to learn. Presently they brought me a delicious cup of tea, with a tortoiseshell cover over it, and a bottle of antique biscuits from Reading; and after a while the lady of the house returned, and I was put with my trunks into her carriage – a sort of big wheelbarrow with a roof over it and no seats, the driver sitting on the shafts. All this was done for nothing, the chief writing to my landlord at Malang that as there was no post-carriage he had sent me in his own. Mr. MacL. received me most kindly. His father had left the Highlands in 1804, and he called himself a Javanese; but in spite of his untidy, disreputable exterior, was a true Scotch gentleman.

He sent me about in his own great open carriage and four horses, first to Singosari, where I saw some huge and hideous old Hindu idols, half human, half animal, carved elaborately out of a stone which is not found in that end of the island, and sitting among palms, ferns, and frangipani trees. The whole neighbourhood of Malang abounds in Hindu ruins, the richest tropical vegetation, running water, and fevers.

The Resident of Malang, M. de Vogel, came and called on me, and looked over my work. He was just like an English gentleman, knew every place and plant, and arranged to send me the next day in his carriage to Djampang with four post-horses, which went like the wind (for which I paid nothing), changing every three or four miles. Around Djampang there was the most glorious scenery – deep dells full of ferns of endless variety, anthurium leaves nearly a yard long, and higher than myself; then endless plantations of coffee-trees, pollards, but growing naturally to the height of twenty feet, they were thirty years old. The best coffee was said to be picked out by a little wild cat or racoon, which eats the fleshy part and leaves the berries on the ground to be picked up and sold. At Djampang many of the fine old forest-trees had been left to shade the coffee; some varieties of Banyan were very curious. We saw a crowd of monkeys in one tree. One of these creatures made a jump which might almost be called a flight. It was a land of jumping or flying creatures – lizards, frogs, foxes, and even spiders flying, or seeming to do so. I saw a huge spider turn and fly at a man

who was trying to catch it. He was not frightened (though it was said to be poisonous), but got hold of all its legs in a bunch behind, so that it could do him no harm.

On my return to Malang, Mr. MacL. arranged that I should go back to Soerabaja by the direct road, fifty miles in a country cart for fifteen guelders; the post would have cost eighty-five, and I preferred this mode of travelling, as I should see more of the country. I enjoyed going slowly and stopping to rest often, when I could sketch the people in the little wayside places; but after the first half of the way my driver got tired (not his horses), and tried to sell me to every carriage he passed, but none of them would take me on as cheaply as he wanted, so he had to go on, grumbling all the way. Having arrived late the previous night at Mrs. S.H.'s, it was nice to wake up in a comfortable room the next morning, and to find the little charcoal-heating machine on the table outside, with its excellent pot of coffee and milk on the top, and pretty china cups. But the weather was too hot for enjoyment down by the sea, so I took the next steamer and returned to Samarang, where the captain was good enough to land me and put me in a carriage, telling it to take me to M'Neil's. Instead of to the bank, it took me to the manager's house – a splendid villa with marble floors, and Japanese pots of roses and carnations all round the verandah. A nice English nurse came out and told me master was getting up and would soon come. I felt quite sorry to spoil his Sunday's rest. He was most kind, wrote me letters, and put me into a nice cool room to wait till it was time to catch the train, and sent me in a good breakfast after my bath. So I got out of Samarang before the mosquitoes even knew I was in it, and reached Solo or Soerakarte at sunset by a slow train, which took me through a rather desolate tract of country with burning forests, showing plainly we were out of Dutch rule and order.

I found quarters in a little mat inn close to the station, and the next morning had two hours' drive about the city, and satisfied myself I did not care to see more, the Emperor and his 999 wives included. I called on the Resident, who said "Oh

Foliage, flowers and fruit of the Sacred Lotus (*Nelumbo nucifera* = *Nelumbian speciosum*), painted in Java.

yes, Prambanan was well worth seeing," and he would give me a letter to the Assistant Resident at Klaten, who could easily take me there. The ruins at Prambanan were more curious than beautiful, with many colossal figures of the gods, the same as those I afterwards saw in India. I was then driven to the station, to wait three hours, to the great enjoyment of its poor Tyrolean master, who seldom got a chance of talking his native tongue to one who knew Meran, his beloved Vaterstadt. While I was gossiping in the room of the Tyrolese, the chief came in with his gilt umbrella of state and followers, one of them bringing a teapot, cup, and sugar under a cloth for me. They all squatted round us, and would not go till they had seen me into the train and off for Djocia – the biggest town of Java, the residence of its native Sultan, and a great stumblingblock to Dutch order in the island. But every one said he would soon be bought off and pensioned. The great square in front of his palace is surrounded by big trees cut like umbrellas, the symbol of greatness in those parts; and a huge elephant is chained up at one corner.

I left Djocia in a grand post-carriage and four, with two extra horses to drag me up the hills, and ten men waiting to haul and push me over the dried-up river beds and lava streams (for grand volcanoes were on all sides). We crossed a most primitive ferry on a great bamboo-mat floor, laid over two boats, with men in hats as big as targets, pulling the thing over by two ropes made of rattan of enormous length. The horses were taken out, the carriage taken down and dragged up from the ferry by men. It was a most lively spot, always full of people going and coming, and animals standing or swimming in the cool clear water. Soon after passing it, we came to a huge cotton-tree, which had nearly strangled and swallowed up an exquisite little temple. Two sides of it were hidden entirely by the roots, between which the poor, crushed, but finely-carved stones peeped out. It was the tallest tree in all the country round, and towered up twice as high as the cocoa-nut plantations near it. The stem must have been quite a hundred feet high before it developed any branches.

About a mile beyond the giant tree and tiny temple we came to the great pyramid or monastery of Boro-Bodo, or Buddoer. At its foot an avenue of tall kanari trees and statues of Buddha

lead up to a pattern little mat rest-house, and the farmhouse of its manager. The house contained a central feeding-room and three small bedrooms. From the front verandah we had a good view of the magnificent pile of building, a perfect museum, containing the whole history of Buddha in a series of basso-relievos, lining seven terraces round the stone-covered hills, which, if stretched out consecutively, would cover three miles. From the top terraces was the very finest view I ever saw: a vast plain, covered with the richest cultivation – rice, indigo, corn, mandioca, tea, and tobacco, with the one giant cotton-tree rising above everything else, and groves of cocoa-nuts dotted all over it, under which the great population hid their neat little villages of small thatched baskets. Three magnificent volcanoes arose out of it, with grand sweeping curves and angles, besides many other ragged-edged mountains. Every turn gave one fresh pictures; and if Boro-Bodo were not there I should still think it one of the finest landscapes I ever saw.

I had the place all to myself, and the good farmer and his son gave me all sorts of good things to eat, all on one plate. I never had any idea what they were made of.

I had nearly exhausted my purse when I got to Magelang, but I had a letter of credit on the landlord of the hotel there. Magelang is a large place, the capital of Kadoe, with the usual central square of banyan trees. Every one was most kind, and the Resident asked me to come and stay; but I did not wish to linger there.

Mr. van Baak wrote a letter for me to the Resident of Wonosobo when he found I was determined to visit the Diëng; and my landlord sent on a horse to the foot of the pass to which I drove. There the chief as usual made me welcome, introducing me to his principal wife, a nice sensible old lady. It was a long drag the next day and nearly sunset before I mounted the hill of Wonosobo – a perfect marvel of richness, and a

Above: The Papandayang Volcano, Java, seen from Herr Hölle's tea plantations.

Below: Mat houses at Bendoeng, Java, with palms and *Brugmansia arborea* in the left foreground.

great contrast to the bare hills we had crossed. On them, however, I found one gem – a perfectly green orchid, and looked forward to a day's rest in a comfortable house, and time to paint it. But there was no hotel, and the Residency was being painted, the family in Europe. However, after some delay the Resident appeared, a singularly nervous man but very good, and before we parted next morning he had become quite hospitable in his offers that I should remain or return, and wrote me many elaborate directions for my future proceedings.

At Garoeng I gave my letter to a most practical and gentlemanly chief, who wore a very stiff stick-up collar and cuffs under his jacket. He and his Head Man accompanied me to the next chief who gave me more tea and biscuits, and a state umbrella carried by his Head Man to accompany me on further. Strawberries were flowering in his garden, and cinchona growing over them. I passed fields of tea full of flowers. The road got always steeper and my beast lazier, and I walked all the last part of the way. The scenery grew very wild, like the top of the St. Gothard; then the plants became like those of Europe (except the tree-ferns, ground orchids, and hollyhocks). At last I reached the rest-house and small village of the Diëng, 6000 feet above the sea, on a small filled-up crater, a pass between the tops of two mountains. It was so cold that I was delighted to roast myself by a great wood-fire. My bed was against the other side of the chimney, and I was right glad of the blanket I brought up with me. The next morning the chief's cream-coloured pony was brought for me to mount, and absolutely refused to hear of such a thing, turning round and round, kicking, neigh-ing and snorting, so I sent it back and walked. The men all tried to get up, with the same success. It was a funny scene. We all laughed, including the pony! I had a most interesting walk among the scattered ruins of tombs, temples, aqueducts, and foundations of big buildings, whose very use and history are unknown. We passed lovely lakes of different colours, saw the mud springs boiling up, and the coils of smoke from them in all directions, with a strong smell of sulphur. Only certain narrow paths were safe to tread on, the rest being a mere treacherous and broiling crust, which would bear no human weight. It was all rather horrid, and the cold caused me such suffering that I

determined to get down the shortest way to warmth again.

I had a most pleasant ride of eight hours on an excellent horse. The young chief, after running round and round the cream-coloured pony for a quarter of an hour, succeeded in mounting him, and rode with me, while my pretty horse went as quietly as a lamb. Java ponies have a habit of resisting their riders' getting on their backs, and showing fight at first, but are excellent and untirable after they are once started. The views were magnificent. We had a long mountain-pass to cross, and much bare moorland. At one place we passed acres of tea in flower with cinchona amongst it. From here we went on to the Resident's big house at Temanggoeng, where two most dear ladies covered me with kindness. I spent four delightful days there, and had a huge garden-room all to myself.

My journey on to Amberawa was a difficult one, the road very bad and horses worse; one poor thing lay down five times, and at last had to be tied up to a tree and left behind. My host at Amberawa said if I would stay longer he would show me many curious things, but I went on the next day to Samarang, thence by steamer back to Batavia, and thence up to my old quarters at Buitenzorg for a few days' rest; after which I took a country carriage with three horses, with extra men to push it when necessary up the very steepest hills, and walked myself up most of the splendid road over the Megamendoeng Pass. Near the top is a deep black lake in an old crater which I went down some steps to see. The large-leaved ferns and arrow-headed leaves of different sorts were most magnificent. Then we descended considerably to Sindang Sari, where there was a kind of hotel and hospital for soldiers managed by an old and somewhat eccentric Dr. Plum.

The garden was full of foreground studies – ferns, aralias, daturas, and areca-palms growing in a half-wild and most picturesque way among rocks and running water, with delicious baths large enough to swim in, through which the water ran in and out continually; beyond all were the grand forests and great volcanoes of Gedé. Under it, about four miles off, was a branch of the Botanical Gardens about 5000 feet above the sea, where the director had a bungalow and spent the summer. The aralias and pandanus were most

Borneo and Java

elegant, and there were masses of a large cane-like plant with a creeping root, called the "patjuy," which produces great bulb-like shoots from the root, of the most beautiful carmine tint, having scarlet flowers and fruits hidden inside. These resemble miniature cobs of Indian corn, full of refreshing juice. They are quite treasures to thirsty travellers.

From the doctor's, six hours in a country cart took me to Tanchur, where the Assistant Resident and his nice family took me in. I stayed there two days and painted the vanilla, with its lovely greenish-white orchidaceous flower-pods and fleshy leaves. I had not seen it in full beauty before. The Resident of Bendoeng picked me up there, and took me on with him in a grand carriage with six horses. He had ruled the Preanger for eighteen years, the highest and largest province of Java, and was a very great man indeed.

We went like the wind. We crossed the river twice, going down almost perpendicular roads and up again, and were dragged by men over a most picturesque ferry; but a new road and bridge were making, and in ten years there would probably be a railway too. We also crossed a range of chalk hills covered with woods in their autumn tints, which might have been in England. The people got more and more civilised, and the chiefs wore black alpaca suits like Europeans, all but their heads, which were still neatly turbaned over their knot of back hair and its comb. I was sorry to see the European dress creeping in; it never looks dignified on an Asiatic.

We passed, at the top of a pass, a lake quite full of huge long-stalked pink lotus (*Nelumbium*) in full flower, a glorious sight; but the weather was so uncertain that the Resident advised my going on the next day while it lasted tolerably fine, and staying with him on my return; so he started me himself and gave me breakfast at five the next morning, packing my trunks into the carriage with his own hands. I was on my way to the house of Herr Hölle, who lived on the hills behind Garoet, and I had a letter to the native Prince or Regent there to send me on.

I had almost made up my mind to wait no longer but return to Bendoeng, when Herr Hölle walked in, a grand man with a strong look that reminded me of Garibaldi, the same curious mixture of simplicity, power, and gentleness. The Governor-General called him "our great civiliser." He drove me in his little single-seated carriage with two small spirited ponies, through Garoet and up a zigzag narrow road, 2000 feet above it, to his village and pretty little house, a model place in every way, ornamented with carved wood and terra-cotta mouldings, all made by natives under his directions. No scene could be more picturesque than those hills crowded with gaily-dressed people amongst the tea-bushes, the plain of golden rice and palm-groves below, with grand mountains beyond, two of them always smoking. I did plenty of painting, but my chief delight was in hearing my host talk, and seeing him among his people. One evening he took me to see the children shaking the trees to collect cockchafers, which they roasted and ate with their rice.

After leaving Herr Hölle I was sent on alone, stopping to shake hands with my kind friends at Trogan, – to Bendoeng, where I found my magnificent room kept for me still, and a kind welcome from the energetic Resident, Mr Pahut. His wife was still at Buitenzorg with her sick child, so I had all my days to myself, and painted a study of the rice-harvest, which was going on all over that rich high plain on which the city stands. It was a bright scene, with the golden stacks, sheds, and stubble, in which the gaily-clothed people and hideous buffaloes were buried up to their knees, with glorious sunshine over it all.

The Resident arranged to send my trunks back by post; he stuck the labels on with his own hands, then packed me off in a great open carriage with Herr von Müller, Head of the "Woods and Forests" of Java.

Travelling with great Javan officials almost takes one's breath away. We seemed perpetually trying to catch some phantom train; horses were waiting at every station, buffaloes at every hill, men running like furies beside the horses, shouting, whipping, pushing, and hauling; people and animals rushing into ditches to make way and show respect.

CHAPTER VII

Ceylon and Home

1876–77

AFTER A FEW DAYS I returned, to pack up and take leave of my friends at Buitenzorg. I was rather limp myself and wanted rest and home, so I gave up my idea of going to the Moluccas, and went back to Singapore in the same steamer which took me to Java. After three days in Singapore, I started in the great French ship *Amazon*, with a good cabin but unpleasant people. The Dutch passengers sulked by themselves at one table, the Chinese at another. I was put among a mixed lot of Britishers, and never spoke a word for four days. There was a good deal of sea too off Sumatra. At last a wild Irishman, who had been wandering all over Australia and New Zealand with his eyes and ears open, took compassion on me and landed me and my trunks at Galle, after which he went on to pass the winter on the Nile and "see if there was anything to shoot there." What a killing race the British are!

The Oriental Hotel at Galle is famous all over the world. Mrs. Barker, the landlady, made me most comfortable, sending all my meals into my room, and I fixed on a "garry" driver I liked, and had him every morning to drive me out. I do not think I knew what cocoa-nuts were till I saw those at Ceylon; there they are the weed of weeds, and grow on the actual sea-sand. The sand was most golden, and the tropical crabs ran over it like express trains. There were also lovely rocks of rich red and golden tints scattered about in front of the

Roadside scene under Coconut Trees (*Cocos nucifera*), at Galle, Ceylon.

sea, and the edge of the sand was bordered with the beautiful sea-grape (as it was in Jamaica), with masses of pandanus on their stilted roots. The sea-waves were exquisitely coloured and clear. I screamed with delight at the sight of a bright green chameleon with a long tail and scarlet comb which ran over the rocks near. My driver made a noose out of a palm-leaf and caught it for me, but the creature's scarlet comb changed to green, and he wriggled so much that I let him go again. It was quite a different creature from those of the Mediterranean.

After eight days of slow stewing, I started in an open carriage (the coach) for Colombo with two young Oxford men for companions, thoroughly nice fellows, just come from China and Japan. The road was most interesting all the way, near the beautiful shore or through swamps full of pandanus and other strange plants, with perpetual villages. I much missed the neat mat and bamboo houses of Java. In Ceylon they were mud-hovels, and everything was less neat, the people lazier, but the little bullock-carts were very pretty. There were plenty of flowers, many of those I remembered having seen in Jamaica.

Colombo is most unattractive, but cooler than Galle. All its houses seemed in process of being either blown up or pulled down. My hotel had "temporary" actually printed on the bills. I sent in my letters to the Governor, and he wrote me a kind note asking me to breakfast, and offering me all kinds of hospitality, but I was anxious to get up to Kandy, and Colombo did not attract me; so he gave me some more letters and sent me off in his own "garry" to the station, ordering a carriage to be reserved for me. Sir William Gregory had a mongoose brought up to show me, which ate buttered toast and snake, killing the latter in the most clever way, springing on the backs of their necks, pinning them down, strangling them and never getting bitten itself. I have never heard any confirmation of the curious story of mongeese combining, one to amuse the snake while another killed it.

It was dark and raining hard when I reached Kandy, and I scrambled into one of the clumsy covered Irish cars of the country, beside a native in a red turban. He turned me out at the hotel. The Governor had told me Mr. Thwaites was going to Colombo to stay with him the next day, so I ordered a carriage at six, and drove over to the Botanic Gardens to catch him before he went. I found the dear old gentleman delighted to see me; and, in spite of the drizzling rain, we had a charming walk round the gardens for two hours. He had planted half the trees himself, and had seldom been out of it for forty years, steadily refusing to cut vistas, or make riband-borders and other inventions of the modern gardener. The trees were massed together most picturesquely, with creepers growing over them in a natural and enchanting tangle. The bamboos were the finest I ever saw, particularly those of Abyssinia, a tall green variety 60 or 100 feet high. The river wound all round the garden, making it one of the choicest spots on earth. Mr. Thwaites showed me also his exquisite collection of butterflies, and promised to give me some of his spare ones. He kept that promise most generously; he never said anything he didn't mean, and detested everything false. He was one of the most perfect gentlemen I have ever known, and I longed to be able to stay a while to rest and paint near him and his beautiful garden. As I was taking leave, I pulled a letter from my pocket, and asked if he knew Mr. L., to whom it was written, and if it was worth my while to give it? He said, Oh yes, he was his best and nearest neighbour, whom he always called the "Good Samaritan"; that I had better go and see him at once, as he was sure to be at home on Sunday morning.

So I turned down a pretty lane, and in five minutes found myself in the garden of Judge L., where his Worship was hard at work, digging in his shirt-sleeves, far too grimy to shake hands, but intensely hospitable. He made me promise at once to move my things and take up my quarters in his spare rooms, in the most perfect peace and quiet, close to the gardens and their good old director, and three miles from the gossip and "Kleinstädterei" of Kandy: it was the very nest I had been longing for. My host was the most hospitable of men. Before he left for India, leaving me in charge, he had friends to dinner most days; but it required a good deal of diplomacy out there to arrange pleasant parties, as many of the nearest neighbours were not on speaking terms with one another. So I saw as little as I could help of all these charming people, and kept quietly at Peradeniya, working either in my own garden or the Botanical close by.

From my window, I could see a thick-stemmed bush (almost a tree) of the golden-leaved croton, and many pink dracaenas, while under them were white roses as lovely as any at home. Then came the lawn, and a great Jack-tree with its huge fruit (two feet long when ripe) hanging directly from the trunk, and branches with shining leaves like those of the magnolia. A cocoa-nut was beside it – a delicious contrast, with its feathery head, masses of gold-brown fruit, and ivory flowers, like gigantic egret-plumes. A thick-leaved Gourka-tree stood also on the lawn, loaded with golden apples, but all hidden away under the leaves out of sight. The lawn was bordered by a hedge entirely covered with the blue thunbergia, hiding the road, along which great bullock-carts were constantly passing, drawn by splendid beasts with humps conveniently placed for supporting the cross-poles by which they dragged their loads.

On the other side of the road was an untidy bit of nearly level ground covered with mandioca (from which tapioca and cassava are made), looking very much like our hemp-plant; bananas, daturas, sunflowers, gorgeous weeds which much offended the tidy eyes of my absent host, but delighted me; a lovely white passion-flower ran all over it, as well as many kinds of lantana, a plant originally introduced from the Mauritius, now all

Foliage and flowers of a Red Cotton Tree (*Bombax malabaricum*) and a pair of Common Paradise Flycatchers (*Terpsiphone paradisi*), Ceylon.

A Vision of Eden

An avenue of massively buttressed Indian Rubber Trees (*Ficus elastica*) at Peradeniya, Ceylon.

over the tropics, and of every possible colour. Pretty hills of about 800 feet surrounded the wide valley covered with scrubby trees; but all looked on a small scale after Java. There was a noble avenue of india-rubber trees at the entrance to the great gardens, with their long tangled roots creeping over the outside of the ground, and huge supports growing down into it from their heavy branches. Every way I looked at those trees they were magnificent. Beyond them one came to groups of different sorts of palm-trees, with one giant "taliput" in full flower. I settled myself to make a study of it, and of the six men with loaded clubs who were grinding down the stones in the roadway while they sang a kind of monotonous chant, at the end of each verse lifting up their clubs and letting them fall with a thud.

Sir William Gregory was the only person the old director ever went to stay with. We both went to dine and sleep at the Pavilion the only night the Governor was in Kandy. The gardens were fine, but the house, from long disuse, looked very comfortless, as its master had not cared to live there since his wife's death. I was put into the huge state-rooms the Prince of Wales had occupied last. His Excellency showed me in, and looked himself to see if they had put my sheets on the bed, for nobody was there to be responsible but the gardener. I felt like a sparrow who had by a mistake got into an eagle's nest, it was such a monstrous place, with one of those odd bunches of flowers gardeners make all over the world, on the table – a dahlia in the middle surrounded by gardenias, then marigolds, geraniums, roses, and heliotrope.

The next morning at six I was at work on my sketch of the outside of the temple, and breakfasted in the old palace, when a party of Indian pedlars came and spread out their gorgeous shawls and other goods on the verandah. They made a fine foreground to the flowers and palm-trees beyond. When I got home I found at last a ripe Jack-fruit to finish my painting from. Denis, the butler, had been constantly looking up at the tree and promising me one "the day after to-morrow" ever since I came, and that one always disappeared and another was looked at with the same answer. Mr. L.'s fruit was always going to be ripe "the day after to-morrow." Though the natives did not steal spoons, fruit was considered common property.

The Governor was coming to open the new waterworks, and a great fête was to be given in his honour. Thirteen elephants were collected to make a show. Some of these had been employed on the work itself, and I was told, carried the great squared stones with their trunks and pushed them into their places, then made two steps back, took a good look to see if they were straight, came and gave a few more pushes, took another look, and were not satisfied unless the work were done with the greatest neatness.

After this my old friend and I went down with the Governor in his special express to Colombo, where I again had the Prince of Wales's great empty room, and after a few days in that dreary grandeur I said good-bye to my kind friends, and went on to stay with Mrs. Cameron at Kalutara. I had long known her glorious photographs, but had never met her. She had sent me many warm invitations to come when she heard I was in Ceylon. Her husband had filled a high office under Macaulay in India, but since then for ten years he had never moved from his room. At last she made up her mind to go and live near her sons in Ceylon and her husband was soon active again. Their house stood on a small hill, jutting out into the great river which ran into the sea a quarter of a mile below the house. It was surrounded by cocoa-nuts, casuarinas, mangoes, and bread-fruit trees; tame rabbits, squirrels, and mainah-birds ran in and out without the slightest fear, while a beautiful tame stag guarded the entrance; monkeys with gray whiskers, and all sorts of fowls, were outside.

The walls of the rooms were covered with magnificent photographs; others were tumbling about the tables, chairs, and floors, with quantities of damp books, all untidy and picturesque; the lady herself with a lace veil on her head and flowing draperies. Her oddities were most refreshing, after the "don't care" people I usually meet in tropical countries. She made up her mind at once she would photograph me, and for three days she kept herself in a fever of excitement about it, but the results have not been approved of at home since. She dressed me up in flowing draperies of cashmere wool, let down my hair, and made me stand with spiky cocoa-nut branches running into my head, the noonday sun's rays dodging my eyes between the leaves as the slight breeze moved them, and told me to look perfectly natural (with a thermometer standing at 96°)! Then she tried me with a background of breadfruit leaves and fruit, nailed flat against a window shutter, and told *them* to look natural, but both failed; and though she wasted twelve plates, and an enormous amount of trouble, it was all in vain, she could only get a perfectly uninteresting and commonplace person on her glasses, which refused to flatter.

While I was at Kalutara I saw the first live snake I had seen in Ceylon. I left my sketching chair under the trees when I went in to breakfast one morning, and on my return saw a beautiful bright-green thing on the back of it waving in the wind. My spectacles not being on, I thought some one had put down some new grass or plant for me, and put out my hand to take it, when it darted off and was lost, and "I did not remain!" It was a riband-snake from the branches of the trees, said to be poisonous. Since that day I have always worn spectacles, and have seen no more live snakes.

I left Kalutara in the midnight of the 21st of January 1877, the whole family going down the hill to the Judge's house with me to wait till the coach came. After a day's rest at Galle I went on board the *Scindh*, a splendid French steamer, on the 24th, which brought me to Aden by the last day of the month, and to Naples on the 11th of February.

From Naples I went by train to Rome where I stayed for three days. I then had two days' delicious sunshine in General MacMurdo's lovely garden at Alassio, then to see Mr. Lear at San Remo with his cosmopolitan gallery of sketches, my brother-in-law, Sir J.K.S., and my niece at Cannes. I went straight through from Cannes to London in thirty-six hours, arriving at midnight

on the 25th of February 1877. After which I enjoyed six months with my friends in London and in the country, the chief event being a visit the Emperor of Brazil paid to my flat at eleven o'clock on the 20th of June, when he looked at all my curiosities and paintings, and told me about my different friends in his country, forgetting nobody that he thought I was interested in, with his marvellous memory. Another event was Kensington Museum sending The M'Leod and Mr. Thompson to look at my different paintings, asking me to lend them for exhibition in one of their galleries. Of course I was only too happy that they thought them worth the trouble of framing and glazing. I was still more flattered when I heard afterwards that in the cab on the way to my flat, Mr. T. had said to the Laird, "We must get out of this civilly somehow. I know what all these amateur things always are!" but in the cab going back, he said, "We must have those things at any price."

I employed the last few weeks of my stay in England in making a catalogue as well as I could of the 500 studies I lent them, putting in as much general information about the plants as I had time to collect, as I found people in general woefully ignorant of natural history, nine out of ten of the people to whom I showed my drawings thinking that cocoa was made from the cocoa-nut.

A Talipot or Taliput Palm (*Corypha umbraculifera*), near the Botanic Garden, Peradeniya, Ceylon. This species flowers just once, producing a massive inflorescence, and, after fruiting, dies.

CHAPTER VIII

India

1877–79

I LEFT SOUTHAMPTON once more by the *Tagus* on the 10th of September 1877, touched for a few days each at Lisbon, Gibraltar, and Malta, and landed at Galle, in beautiful Ceylon, on the 15th of November, took my passage to Tuticorin by the next steamer, and spent some of the intervening days in visiting old friends in the island. I took a carriage from Pantura and drove ten miles to Kalutara. The road was a series of beautiful pictures all the way. I found Mrs. Cameron much as I left her, the old man even younger and happier. I went to Peredeniya the next day, and up to the Botanic Garden to see my old friend the Director. I had a long stroll with the dear old man, who looked much aged, and so delicate that a touch might have knocked him down. The next morning I started before daylight to catch my steamer, but the heavy rains had broken down the roads, loosening great rocks which blocked the passage entirely. I had good friends who stuck by me all day. We did not reach Colombo till 5 p.m., when I found my steamer had left hours before, so I had another week to wait on the island.

At last I embarked and had a good voyage to Tuticorin, in an excellent ship with pleasant passengers. The Captain did not half like letting me risk the six-miles' row over the sand-breakers in a loaded boat, but at last we started and had a

good sea. We were two hours and a half landing, but got in dry, then had to scramble for rooms at the rest-house as the train had gone. Friends were kind to me as usual, and the next morning they started me with a basket of provisions in the train for Madura.

The first part of my Indian journey was over white sand covered with palmyra- and fan-palms, and cacti; then came cotton, quantities of millet, Indian corn, gram, and other grains. Mr. Thompson met me at the Madura station, and took me home to his comfortable bungalow.

The next day I was quite dumbfounded by the strangeness of the old Temple of Madura. It was full of darkness and uncanniness, with monkeys, elephants, bulls and cows, parrots, and every kind of strange person inside it. The god and goddess lived in dark central stalls to which no unbeliever is allowed entrance; but two small black elephants with illuminated faces, painted fresh in red and white every morning, wearing wreaths of flowers round their necks, were admitted into that "holy of holies," with a youth riding on the head of each, and carrying a silver vase of water. The dignity of that proceeding was tremendous. In these temples there is an endless variety of courts and columns, more grotesque than beautiful, with dragons, griffins, gods and goddesses, larger than life. In the entrance are money-changers, and all sorts of merchandise, a gorgeous variety of bright-coloured cloths with gold borders, which both men and women wrap round them like a petticoat all over India; these are made at Madura. In one part of the temple is a hall of a thousand pillars, all

Pagodas of Mandura, India, in the distance with, in the foreground, foliage and flowers of a Portia Tree (*Thespesia populnea*). The flowers are at first yellow, but change to red with age.

different. I made a sketch of one of the small inner temples, in which the god and goddess were married every year. When we came away, wreaths of sweet trumpet-flowers were put round our necks, and a lime to smell in our hands (a very necessary luxury in such a locality).

I painted a sunset view of the grand tank outside the town, with its island-temple and palm-trees, grand old banyan-trees and other temples on its edge. The English people drive round and round it every evening, and make that drive their chief gossiping spot.

Starvation, floods, and fever were all round. The railway was washed away in nine places, and I could not have left it even if I had wished. The gardens were all flooded, the great river rising, tanks breaking on all sides. The pandanus pines alone seemed to enjoy it, being buried in water up to the tops of their odd stilted roots. Every one was taking opium, so I followed the fashion, prevention being better than cure.

I left in a special carriage ordered for the judge and his wife, and we reached Dindigal at sunset, where Mr. M. met me, and drove me to his charming wife and home. They pressed me to stay and make an excursion to the Palani Hills, 8000 feet above them, which were most lovely; but I could not linger, and at five o'clock I was in a long chair on the shoulders of four coolies, with eight more to relieve them, and two peons to drive them on. At last we reached some waggons, full of coolies, which took me to the train, and so into Trichinopoli, where Colonel F. met me and drove me out to his home at the camp. On the 24th of December 1877 I reached Tanjore by the earliest train, asked a policeman I saw to show me the way to the Doctor's, and walked under his porch about nine o'clock, to his great surprise, as he was sitting among his books deep in work, having expected me by a later train. Living with Dr. Burnell was like living with a live dictionary, and was a delightful change. He had all sorts of sacred Hindu plants ready for me to paint (he having undertaken to write their history at the same time, and to publish it some day with my illustrations). He made me feel quite at home, and in no hurry. He and his friend showed me the splendid temple, lingering over all its rare bits of carving and inscriptions till I felt at home there too. I know no building in its way nobler than that temple of

Tanjore. The colour of its sandstone is particularly beautiful; its whole history is inscribed round the basement in characters as sharply cut as if they were done yesterday. I did one large painting of the outside, driving every afternoon to the point of view I had chosen, where the Princess of Tanjore had ordered a small tent to be put up for me, and a guard of honour to attend me!

The real hot weather came, and to me was enjoyable. I was very sorry to leave Tanjore and its good talk, such as I was little likely to meet for months to come. The F.'s put me up most kindly again, and the next morning the Colonel took me to Seringham, the largest temple I ever saw: a perfect city in itself, but very dirty and rubbishy. The view from the roof was the most curious part of it, and gave one the best idea of its great size. The terminalia trees produced a strange effect, with their rectangular branches and deep-red leaves. We dined with the Judge that night, who said I must also see the Sira Temple near Seringham; he would send a peon to show me the way. So I went, and was glad, for in many ways it was more interesting and picturesque than the other.

Grey's Hotel at Kunur consisted of a number of small bungalows, dotted about on lovely terraces and gardens, round the central boarding-house, where the master and mistress lived, and had their chief kitchens. I had a most luxurious little house all to myself; it was furnished with carpets, fireplaces, four-post bedsteads, and every kind of English luxury and absurdity. An old woman kept the fire going of an evening, and washed my clothes; and a grand man in a turban brought down four great covered dishes twice a day, with tea at seven and three. Every hill was tinged with red blossoms; they were scraggy, shabby trees, not bigger than English apple-trees, the flowers decidedly poor, with a white eye, but of the deepest red. I took some walks through the rain and clouds. Then, as I felt my limbs and ankles beginning to swell and stiffen, I decided to avoid rheumatism by coming down again to a warmer climate, resisting an invitation to go and stay at Utakamund.

At Beypur I found a large room over the station, a hundred yards from the sea, with a garden between me and it. Also a servant engaged for me by a friend of Dr. Burnell, who had come from Cochin on purpose to attend on me. I enjoyed being at Beypur close to the sea, with no dirty town. I could walk on the rocks and sands, watching the shrimps, crabs, and other queer creatures in their own home-circles. I made a long sketch of the river and distant mountains, with endless cocoa-nuts in the middle distance, ferry-boats, and picturesque people. It was very pleasant sitting on the clean sand, but it was hot. The jack-crows were the chief objection to my quarters at Beypur. They flew in at the window and stole every small thing they saw; I caught one just hopping off with a tube of my precious cobalt one day, and only came into the room in time to make him drop it.

I had been waiting some days for the steamer, but suddenly determined, from what I was told to go to Cochin by backwater instead. Cochin was full of Christians and beggars. I went out for a stroll past the old church and Frank settlements and through the Jews' quarter, saw the synagogue which Dr. Burnell said was built in the seventh century, Cochin having been a port to which the old Egyptians used to traffic; later still King Solomon himself sent his ships there.

I started at four in the afternoon in a big cabin boat, with thirteen men and a tame old Moslem as a servant. My crew made a frightful noise all night, singing and rowing furiously. We passed over huge inland seas, rivers, and narrow canals again, and reached Quilon about twelve the next day. I decided to rest the night there in the bungalow, which is a mile from the river, and deliciously airy, surrounded by cashew and mango trees. Thence on to Nevereya, where we left the boat and crossed the boundary in a bullock-cart. We went on in another canoe, hollowed out of one long tree, for twelve hours more, stopping to breakfast at a cocoa-nut farm within a stone's throw of the salt sea on one hand, and the backwater canal on the other.

Trivandrum is a model little capital, buried among tall trees. I stayed there with Dr. Houston, who got me rare plants, and told the men where to take me to see the prettiest views and to sketch. The little toy houses were something between Tyrolese and Arab, with tiny double-arched windows and slender marble shafts, so small that one could not get one's head through them. I met the Maharajah taking his walk one day. He shook

his hand at me, and said, "I hope you are quite well." He always said that to all Europeans.

I returned to Cochin very much done up, and hoped for a few days' rest, but heard that the steamer for Bombay would be in that afternoon, so had to be ready, and got well over the bar at the mouth of the harbour, and on board the very nicest little steamer I was ever in, the *Khandala* of the British India Line. The entrance to Bombay is very striking, with its numerous islands and abundant shipping. I was not allowed to stay at the hotel. Sir Richard Temple's secretary, Major R., sent to ask me to move at once to Government House.

The sunrise every morning from the rocks behind my room was beautiful. It used to come up like a round red ball behind the purple hills and hanging smoke of the city some five miles off, and the red-coated servant used to bring my *chota hazra* or early breakfast out on that rock to me

Male inflorescence of a screw pine *Pandanus tectorius*, with entire trees and buffaloes wallowing in the mud of a swollen Indian river in the background.

A Vision of Eden

every morning. Around me were wild peepul-trees, full of berries; erythrina-trees, with their red flowers just opening, and wild cherries. Below all was the sea, and a perfect fleet of boats with bright sails going off after fish. The first morning Sir Richard Temple gave me a walk before breakfast, showed me all the odd trees, the beautiful stable of horses, and the great stretch of brown rocks, up which a perpetual stream of women came, carrying water-jars on their shoulders, full of salt water for the roads.

Mrs. C. very kindly found me a servant, a Madras man, who called himself John. He had a gorgeous turban, bright black eyes, and most limp long figure. He and I started off by rail for Neral, after a week of luxurious idleness at Government House; and there I mounted a pony and rode up to Matheran, which, Mr. Lear had written me, was "a highly Divine plateau." The views were certainly fine, having strangely shaped rocky hill-tops in the middle distance, with almost vertical strata of different coloured trap rising some 2200 feet from the great plain and distant sea and islands about Bombay.

But a few days were enough at Matheran, and I rode down, took the rail on to Lanawali over a magnificent piece of engineering, up the Bhor Ghat, one of the finest bits of scenery I ever saw. The plain was covered with golden stubble, fine trees dotted about, and stacks of rice and corn were built up on stilts or in the middle of the spreading trees.

I had the house to myself, and thoroughly enjoyed it; but my chief object in coming there was to visit the Cave of Karli. At last we came to the final climb over the hard volcanic rocks, and first to a splendid tree of the *Jonesia asoka*, full of orange flowers and delicate young lilac leaves. The priest of the temple found me one fine flower growing through a honeycomb full of honey, which had been built round its stem. Now this was a very curious thing. Did the buds push their way through the honey and wax, or was the thing built quickly round them? I never satisfied myself which was the first perfected. The cave itself, more interesting than beautiful, is accurately described by Mr. Fergusson, who also gives an engraving of it in his *History of Indian Architecture*.

On the 26th of February I left Bombay at eight in the morning. A railway took me up another splendid pass to Nasik Road, where I transferred myself, my luggage, and John into two low dog-carts, drawn by ponies, with the driver sitting astride on the pole, his feet clasped under it. Our first stop was at Nasik, a most picturesque old town, with steep busy streets gaudily coloured and carved in wood and stone; the river banks lined with temples most beautifully ornamented, and paved with stone, having grand flights of steps, and many causeways across. The next day we reached Aurungabad and the following morning we went on to Daulatabad, the famous Indian fortress. It has 120 feet of sheer perpendicular precipice all round it, and many subterranean passages, stairs, and halls, through which one must pass to get to its top. I saw a tiger-trap outside, which had lately caught its game, a sick kid being the bait. After leaving that horrible fortress, we drove on to another ruined city – Roza, which has some very beautiful mosques, and tombs of kings, with doors of silver and gold, and half-precious stones in the inlaid pavements. A mile farther down the steep road was Ellora, where I found twenty-four caves of every age and variety of design, but all insignificant compared to the great Kylas, which is a perfect cathedral, standing on the backs of some hundred elephants, nearly as large as life, all cut out of the solid rock. I never saw any building so impressive and so strange.

At 2 a.m. I was rolling on again, in one of the comfortable carriages of the G.I.P. Railway, and slept nearly all the way to the old town of Jabalpur. I started at night again for Agra, which I reached on the morning of the 14th of March. The ground all round the city was pure dust – one ate it, breathed it, drank it, slept in it – but the place was so glorious that one forgot the dust entirely. I went that same afternoon to the Taj, and found it bigger and grander even than I had imagined; its marble so pure and polished that no amount of

Above: The Taj Mahal at Agra, northwest India, with in the foreground Poinsettias (*Euphorbia pulcherrima*), bougainvillea, cypresses, with Rattan Palms (*Calamus tenuis*) attached to the tree on the right.

Below: The road up to Naini Tal, India, with a tree of *Desmodium oojeinense* (= *Ougenia dalbergioides*) in a spring-time burst of flower.

dust could defile it; the building is so cleverly raised on its high terrace, half-hidden by gardens on one side, and washed on the other by the great river Jumna. The garden was a dream of beauty; the bougainvillea there far finer than I ever saw it in its native Brazil. The great lilac masses of colour often ran up into the cypress-trees, and the dark shade of the latter made the flowers shine out all the more brightly. The petraea also was dazzling in its masses of blue. Sugar-palms and cocoa-nuts added their graceful feathers and fans, relieving the general roundness of the other trees. The Taj itself was too solid and square a mass of dazzling white to please me (as a picture), except when half hidden in this wonderful garden, though on the river side it was relieved by wings and foundations of red sandstone.

I was very ill, and found it of no use fighting longer with the dry heat of Agra, so started by the rail at nine at night for Bareilly and then on to Kanibagh in the district of Kamaun, a place with large mango-trees all loaded with bloom, and a running river below, which did one's eyes good to see after the months of dust. I felt better, and could eat again, and after a night's rest was carried up the hills in a dandy (by two bearers at a time, changing places every five minutes), in four and a half hours, to Naini Tal. I never saw such abundance of pure colour; but they said in a few days all that blossom would shake off, and I found it was so. When I returned a few days after to sketch, it was already gone.

After less than a fortnight at Naini Tal, I was a new creature, and able to walk over the hills, which were just then showing their spring foliage. But the snow came down into the valley, so I determined to go down again, to seek my sacred plants in the Sahāranpur gardens. I took one of the returning carriages to Bareilly, then on by Aligarh to Sahāranpur. The next morning I drove to the gardens soon after daylight, and called on Mr. Duthie, almost before he was dressed; but he soon came down, and walked about the gardens with me. He found out the trees I wanted, few of which were yet in flower, but he said he would let me know in time to come down and paint them, even if I were up in the hills. He had expected me for some time, and had arranged with Doctor and Mrs. J. to take me in when I came. I very soon knew every one in Sahāranpur, but I did not stay

long, as my plants were still some way from flowering.

On the 24th of April I started in a dawk carriage (a heavy wooden close fly) for Rurki, where we crossed a fine bridge, guarded by two splendid stone lions, over the Ganges Canal, which is one of the grandest pieces of engineering in the whole world – not a sluggish ditch, but a rushing snow-fed river. The next morning Major T., the head of the canal works, called on me in his dog-cart, and drove me up the edge of the canal to Hardwar. The town was a perfect museum of rare old buildings, marvellously carved and painted. On one wall was represented the taking of some city by the English, who fired off cannon like pistols, and had a brandy-bottle under the other arm. It is a most enjoyable place for an artist, full of pictur-esque bits of street views.

At Dehra, where I stayed four days, I painted some more of the sacred plants, which I caught in flower. Dehra is famous for its bamboos, which, however, had followed the fashion and flowered the year before, so looked at their worst when I was there, though they were throwing up fresh canes from the roots. The old ones were very shabby, dead, or dying. It was the same year that all the English bamboos flowered and died, as well as those in Spain and France. In India they only died down and started afresh.

I drove off to the foot of the hills, then was carried up the zigzags to Masūri, which is a long scattered place, covering an uneven ridge for about three miles, looking over the wide Dūn valley on one side, and into the rolling sea of mountains on the other. The nearer mountains were topped with pines, terraced with burnt-up corn and grass crops. All the rest was bare, except some narrow strips of green following the watercourses. Over all was a zigzag line of snow-tops, full 26,000 feet above the sea, with no one peak particularly domineering. The idea of their great height came from seeing them over so many ranges of mountains, themselves so very high and close, they would certainly have hidden any European line of snow peaks (such as that which

The great lily of Naini Tal, India; *Lilium wallichianum* grows to 6ft or 7ft high and is seen here associated with *Chirita urticaefolia* and a *Begonia* species.

A Vision of Eden

one sees from Berne for instance). After my arrival Colonel and Mrs. W. came to see me, and insisted on my leaving the hotel and going to stay with them; and I spent a most delightful fortnight in their pretty house.

Mrs. W. took the greatest pains to show me all the most beautiful points in the country. We went everywhere in dandies, like the rest of the Masūri world, with our carrying men dressed in a sort of livery of black native flannel, knicker-bockers, and jackets edged with scarlet, scarlet turbans and sashes. She had five and I had five. Two men at a time balanced the small canoe in which we sat, or rather reclined. They liked to go at a fast trot through the streets, and it was rather nervous work when the bazaars were crowded, for no one would slacken pace or give way to others. The hills were full of wild nooks, great overhanging rocks, and scraggy twisted oaks, with the gay leaves of the Virginian creeper looped about them. They must have been quite dazzling in autumn. There were also masses of white dog-rose, with a much larger flower than ours, and two primulas and an androsace were creeping over all the dry banks. The climate was delicious between the showers, which were as frequent as in England, and I doubt whether England itself could have shown finer roses than grew in the gardens about that place. At last I turned away. Both my friend and myself spent two days at Dehra, with Dr. and Mrs. J.

I went on to Amritsar, where the hotel was not bad. At breakfast one of the officers told me he had been there some months, and never yet had had the curiosity to see the Golden Temple! I went at once, having picked up a garry-driver at the station, who talked some English. It was a real gem, half white marble lace-work, and half gilt copper, with rich dark hangings and carpets, built out in the middle of a clear lake, smooth as glass, in which every line was accurately reflected; a long causeway of marble leading to it was always crowded with finely-dressed people. The lake was surrounded by trees and picturesque buildings. I set to work at once on a sketch, and no one interfered with me the first day, but on the second they said, "No orders give chair," and would not let me sit, even on my own, anywhere. However, I had done most of my work, and did not care.

I soon went on to Lahore where I stayed with the Judge and Mrs. E. in a most delightful garden. The flowers at Lahore were in their most gorgeous state. The Amaltas or Indian laburnum was a perfect mass of yellow, with flowers and pods more than a foot long, and the tanks were full of tall pink lotus. The drives round the old city are most charming with noble trees shading the road, and beautiful peeps through them of the old walls, gates, mosques, and minarets.

All the country round Lahore abounds in fine tombs, and many of the European houses are made out of them – even the Governor's own house and the English church, the grand domes making fine central halls. At the Governor's one was used as a dining-room. Tombs and gates are covered with beautiful encaustic tiles, and one mosque outside the city was quite a botanical study. All the best-known country plants, as well as the iris of Persia, were represented in a kind of tile mosaic, up to the very tops of the minarets, –

Study of a Deodar (*Cedrus deodara*), in full cone and clothed in a creeper, probably *Cissus himalayana*, Simla, India.

India

Deber Dhoora Dee, Kumaon, India, with its well and Deodars
(*Cedrus deodara*).

beautiful turquoise and lapis lazuli, – blue, green, and yellow being the prevailing tints. The streets are full of colour, and the houses of the Sikhs high and square-topped, so that I found incessant ready-made pictures, if I had but had time to paint them. But, after the two first hours of the morning, it got too hot for working out of doors. Mrs. E. seldom went out till after six. Then came the sunset, and the day was over.

I went on to Simla – the Hill Capital of India. When we came to the end of the dawk road, I found men in royal red coats and a jampany waiting to carry me up to the Deputy Governor of the Panjab, where I was most heartily welcomed by Philip Egerton's brother, the great man himself: one of those who work hard and like it. He showed me into my room, sent me some tea, and took pains to make me feel at home. My window looked over endless hills, with great deodara branches, their stems and cones for foreground. Two ayahs followed me in, and fought for the possession of me, though I wanted neither.

I wondered what people had meant when they told me Simla was not beautiful. I found endless subjects to paint close to the Governor's house, and used to slip out before the live bundles in the hall had done sleeping, and shaken themselves up into a sitting posture for the day (which was all the toilet they ever performed). I used as usual to work indoors during the heat, and generally the Governor called me out for a stroll with him in the evening. He knew more about the plants and trees than any one I met in India.

Our house was on the southern side of a mountain-top called Jako. On that side the deodara had it all its own way. On the north the oaks turn it out; and under them there was more variety of vegetation. The benthamia was the most striking thing, with its four primrose-coloured bracts, shaped like the bougainvillea. There was a delicate white-flowered spiraea, which Anglo-Indians called the white thorn, pale pink, wild indigo, many varieties of dog-roses, columbines, delphiniums, primulas, a tiny white lily, orchids, and cypripediums. I walked all over Jako with the Governor, and had a nice walk one day with General and Mrs. S., at the other end of the place. The General was another walking botanical dictionary. Then I went to stay with Sir Andrew and Lady C., at that same end, and saw quite a different set of people in their house. Lady C. delighted in society, and had constant dinner and luncheon parties.

One evening I dined at the Lyells', and met Colonel Colley, who said I ought to go and make some sketches at Nahl Dehrah, and arranged it all for me; and I went off there round Jako, through a tunnel, and on to the viceregal camp, under the splendid old deodaras, which surrounded the quaint little wooden temple of Nahl Dehrah. Below that was the deep valley of the Sutlej, and noble mountains beyond it, leading up to the perpetual snows. The place was like a hive of bees when I arrived. All the grand people were doing "kef" after breakfast under the trees. They had a tray brought back for me. Then I painted all day in a very dusty, dirty state, for there was no tent to be had till they had all departed in the afternoon, when Lady Lytton was to leave me in sole possession of her own. But her manners were as kind and sweet as if I had been as well dressed as herself. It was a pretty sight to see all the party depart, and reminded me of a hunting-field at home, so many red coats among the green meadows and trees. After that it became quite quiet; only a few tent-men and servants were left to wait on me. Lady C. had lent me her mate or head road-man, and young C. his "boy," so I was well guarded, and the Viceroy's last words as he rode off were, "Not to be frightened if I heard noises in the night. There were no robbers – only leopards and bears." But they did not think me worth eating, and all was peace.

I can never forget that sunset and sunrise. The flat-topped old deodaras draped with Virginian creeper made delicious foregrounds; the young green cones looking almost white against the dark foliage and boughs. Great golden eagles came rushing across the deep valley, looking really golden in the slanting sun's rays, against the blue misty mountains. I had to hasten back to dine with Sir John and Lady S., or I could have lingered longer at Nahl Dehrah.

At last the much-longed-for rain had begun.

After a short visit to Mashooba and Fargoo, I returned to Simla and soon went a couple of hundred yards down the hill to Mr. A.C.L.'s, the Foreign Secretary, quite the cleverest man I had seen in India. His talk was full of odd original ideas; but he was much overworked, and often did not talk at all. I liked Mrs. L. very much, too. She at once understood my wish to get to Markunda before the rains rendered it impossible; and that every day was of consequence, – a fact no one else in Simla had believed, they were so accustomed to feminine helplessness.

Mrs. L. at once arranged that I should have a chaprasi from Captain N. and a man of her own to cook, and sent me off. The first day was over bare hills, steep, with some terrace-cultivation on the lower slopes, with yellow burnt-up crops. The road was a grand one, nearly level, over the tops of the hills. It rained and thundered at intervals, and when we left Fargoo one of the coolies tried to escape. They hated getting wet, like cats, having like them only one set of clothes. At last we came to the summit of the pass, – a rude cairn or temple, stuck all over with bits of red cloth and rags, – and descended on the other side fifty feet to the rest-house of Markunda. A perfect sea of cotton-wool clouds surrounded it, and it was very cold; but at sunset they all cleared off, and I had the full view of the great snow range, one of the finest in the world, with some of the highest peaks quite crimson, and the rest in shade.

The ascent was nearly perpendicular in some places, and the eight coolies were quite disappointed that I did not submit to be carried all the way. The ground was blue with forget-me-nots and blue anemones, just like our white wood-anemones. The maiden-hair fern covered the ground between the flowers, as grass does in England. The road beyond Markunda wound along through the grandest forests of pines 200 feet high, often draped to the very tops with Virginian creeper, and the peeps of snow-mountains and blue sky seen through these were enough to drive one wild.

Distant view of Mount Kinchinjunga from Darjeeling, India, with tree ferns and oaks festooned with *Thunbergia coccinea*, in the foreground.

132

A Vision of Eden

Our return journey was made through torrents of rain. It was as much as I could do to keep emptying the pools of water which collected on my mackintosh cover, and every flower which could shut itself up, did. I went down the hill and back to Dr. and Mrs. J.'s at Sahāranpur for a couple of nights, and there I heard, "Oh, you will have to stay here, for the road to Darjeeling has been washed away, and has been impassable for four days." They showed me the paragraph in the paper, and telegraphed to the Calcutta dawk-contractor for the truth, and the answer was, "Both the roads interrupted." It was hot at Sahāranpur in July. Poor Mrs. J. looked washed-out and limp. We were both inclined to sit in rocking-chairs and read novels, and little else, but had a stroll in the beautiful gardens before breakfast.

As I could not get to Darjeeling, I determined to return to the Kamaun district for a month. What a contrast to my last visit as a wretched invalid,

Below: Rhododendron nilagiricum is unusual in that it is native to the Madras region of India, although some authorities believe it to be just a variety of the Himalayan *R. arboreum*.

hardly able to crawl! I could scarcely make up my mind to come down. Then I went to see Mr. G., the Head of all the Forests, and his wife. He lent me his map, and sent me with a note to the magistrate to beg the loan of a chaprasi, an official who wore a band which gave him a certain authority over the natives everywhere. All travellers have a right to the services of such a man. He was better than any other servant, and only cost six rupees a month. He found me ten coolies to carry myself and all my goods, and the landlord lent me an old dandy. Deber Dhoora Dee was the object of my pilgrimage. It was a very singular place – a nest of poor little temples and great granite boulders. There were six noble deodaras with flat tops like cedars, but they were not finer than those of Nahl Dehrah and many of the large ones near Simla. A fair was still going on, chiefly of Manchester cottons, and the pavement near the chief shrine was still covered with the blood of buffaloes sacrificed to Shiva. Goats were also killed and eaten, after being roasted whole, the head and feet being given to the priests. All those ceremonies are very like the old Jewish ones.

We returned to Naini Tal, descending among its villas and civilised ways to be told that it had rained incessantly ever since we left it a month before. I was nearly at the end of my money, so I went to the bank (there was one on Coutts' list), showed my letter, and asked an old gentleman there if he would give me some money. He read the letter of credit slowly three times, and said deliberately, "It cannot be done." "How funny. Why not?" said I. "Because it is a most irregular proceeding." So I wished him "good-morning," and went up to Mr. G. and told him. He said, "Like him, the old idiot!" and lent me fifty rupees tied up in a handkerchief; and the next day I descended to Karledone in torrents of rain (I was not sorry to leave it all), and was dragged on all day and night in a garry till past twelve, when the road came to an end in the river opposite Muridabad.

The bridge was washed away; there was water in front and on both sides close up to the road, and several garries were waiting there, with the buffaloes half in the water. My man unharnessed

Opposite: Foliage and flowers of two Indian rhododendrons – *Rhododendron griffithianum* and *R. arboreum*, the commonest mountain species.

the horses, and let them trot back again to their last post-house. There was no boat, and nothing for it but to shut oneself up in that box on wheels and wait till morning. Storms of rain, thunder, and lightning came on, and when dawn broke I found only a few inches of road above water. The river was rising steadily, and I watched the last blades of grass disappear under it at the edge of the road. At last the waters closed under the wheels. The natives, apparently content to wait the possibility of dry weather and of the bridge mending, sat under mats on the tops of the garries, and would have gone on so for a week. I was soaking, and so was everything else. Then I discovered one old boat, full of water, and ordered it to be emptied, getting into one of those rages which are sometimes necessary when dealing with semi-savages, and fearing all the time the boat had a hole in its bottom, but it had not. Two men set to work with a bucket made of matting, slung between them on ropes. Themselves standing on the seats, they slung the bucket cleverly in and out, and in about an hour the boat was empty. All the other wrecked travellers begged a passage in it. They stuck up a kind of thatched fan to shelter me from the rain, and moved in the luggage.

We started with my shivering coachman and guard – the latter a terrible personage with a big sword, and his upper lip cut in two like a bull-dog; but he was most kind and careful of me. We had to go a long way round by the lee side of islands, and the men pulled and hauled to keep the boat out of the great currents. At last they landed it below the town, on a mud-bank, and we scrambled on shore, finding our way over the slippery banks and lanes to a kind of square or open place. Here we rested ourselves and our things on the wall of a well to wait for a garry which the guard had sent for; when a tidy official appeared and coolly demanded toll for the bridge and barrier which did not exist! I refused, and laughed in his face. He insisted, but at last laughed too, and brought a slate and pencil for me to write my reasons for refusing, which I did, and have little doubt "the sahib" also laughed when he read them, for I heard no more of such myths as bridges and barriers. The hotel I expected to find was shut up, so I drove to the railway-station, and after much search a key was found to the ladies' room. The door was opened with a good thump, and I found a comfortable

dressing-room and a most kind station-master, who sent me an ayah to dry my things, saying, "he saw everything I possessed was mortal wet," and he told the Khansamer to feed me.

From Muridabad, I passed through Lucknow and Benares to Calcutta, where I soon found my way to the huge cosmopolitan hotel. The famous botanic gardens are six miles from Calcutta, but the whole drive is full of interest and wonderful vegetation. A German was director of the gardens in Dr. King's absence, and went heart and soul into my work of hunting up the Sacred Plants. He put me into the hands of a learned baboo who said "it pleased him much that I should take so much trouble about the flowers that Siva loved," and he told me many things about them. One plant, the "Bah," a famous cure for dysentery, he said he never passed without bowing to, and always put a leaf in his pocket every morning, then nothing could happen to him, – he must be safe, as Siva loved all who were near these trees. He also told me that when he felt old age coming he should go to Benares and die there, and so be quite sure of going to heaven.

The flowers I was in search of were still out of bloom, so I left Calcutta again the next morning at 7.30, and soon had to change trains, then got into a steamer and crossed the great Ganges river, then into another overcrowded train. The plain of Bengal is wonderfully rich, full of sugar, Indian corn, indigo, arrowroot, hemp palms (toddy), and bamboos, in magnificent luxuriance. The cottages of the people were neater than any I had seen before; they are built like those of Java, on stilts, made of neat matting and bamboo frames instead of mud, the roofs rounded, which gives them a beehive look.

The next day took me over the most glorious road, among forests and mountains, to Darjeeling, the finest hill place in the whole world; and I brought my usual luck with me, for Kinchinjanga uncovered himself regularly every day for three hours after sunrise during the first week of my stay, and I did not let the time be wasted, but worked very hard. I had never seen so complete a mountain, with its two supporters, one on each side. It formed the most graceful snow curves, and no painting could give an idea of its size. The best way seemed to me to be to attempt no middle distance, but merely foreground and blue mistiness of mountain over mountain. The foregrounds were most lovely: ferns, rattans, and trees festooned and covered with creepers, also picturesque villages and huts.

The people, too, were unlike any I had seen before, – natives of Butan and Thibet, who come every year to make money, during the season. They are rather like Chinese, with flat faces, long eyes, and long hair. They are intensely good-humoured, laughing and singing, very industrious and strong. The flowers about Darjeeling seemed endless. I found new ones every day. The *Thunbergia coccinea* was perhaps the most striking; it twined itself up to the tops of the oaks, and hung down in long tresses of brilliant colour, the oak itself having leaves like the sweet chestnut, and great acorns as big as apricots almost hidden in their cups. There was another lovely creeper peculiar to Darjeeling, – the sweet-scented cluster ipomoea, of a pure pink or lilac colour. The wild hydrangea with its tricolour blooms was also much more beautiful than the tame one. I worked so hard and walked so much that, after a dinner or two with Sir Ashley Eden and other grandees, I refused any more invitations. I could not keep awake in the evening.

How I longed to spend a spring in Darjeeling, and to see all the wonderful rhododendrons and magnolias in flower! They were such great old trees there, and of so many different varieties. One hairy magnolia was then in flower, and the Lieutenant-Governor had a branch cut down for me one day; it was very sweet. Major L., the head of the police, was kind enough to lend me his interpreter – a most grave and responsible character, with long eyes and a pigtail. He wore a brown dressing-gown and a Chinese cap turned up like a beef-eater, and whenever I made the slightest remark, he got out his note-book and made a memorandum of it, like an M.P. in search of facts. His name was Laddie. He spoke excellent English, and declared it necessary for me to have twelve coolies and a cook to go about the hills with me, bringing up my expenses to over thirty shillings a day. I started in a dandy the Governor lent me, carried by a fine band of ragamuffins; it was almost worth the extra pay to see their picturesque figures and independent air of insolence.

It was late in the morning before we got started,

Valley of ferns near Rungaroon, India.

and ascended the steep hill through the camp, when the young officers darted after me, "Where are you going to now, Miss North?" for they all knew me by sight, although I did not know them. Much of our way led through the forest. At one place a great fallen tree filled up the path, covered with rhododendrons and other parasites, and I saw my old friend the aralia again. Rangerom is a mere clearing in the forest on the steep hillside, which some insane governor had once made to grow cinchona in. It was then turned into a botanic garden of native plants and pines, and a poor Scotch gardener was slowly dwindling away with fever and ague among them.

A Vision of Eden

While hard at work at that fairy dell I felt it was raining, and before I could get over the fifty yards of steep descent to the bungalow with my things, I was soaked through and through, and came back through a running stream of water to find the house occupied by a large picnic-party – a regular ball-supper, cooks, coolies, and other litter all over the passage floor, and half a dozen ladies all drying their things and themselves in my room, using my towel and soap, almost too much company to be pleasant. I escaped as soon as I could to my poor soaked painting. "You only sketch it on the spot and paint it indoors?" one beauty said, pointing to the poor thing which was so covered with raindrops that it looked as if it had the smallpox. "Yes," I said, "that's what I do. Then I take it out to be rained on, which makes the colours run faster, and that's why I paint, as you say, so quickly." Those unthinking, croqueting-badminton young ladies always aggravated me, and I could hardly be civil to them. I had not met a single person at Darjeeling who had seen the great mountain at sunrise, and few of them had seen it at all that year. Kinchinjanga did not keep fashionable hours.

From the hill above Jonboo one saw the plains of Bengal like a sea, and mountains on the other three sides. The clouds rolling in and out of the valleys and up into the sky at sunset, quite took one's breath away with their beauty and colours. It was then too cold for tents. Frost was white on the ground round me when I began, at sunrise, to paint the highest mountain in the world – Deodunga, or Mount Everest as it is now called. It forms quite a distinct group, detached from Kinchinjanga by a hundred miles at least, and its form is much less graceful and definite. The view was perfectly clear for two hours; all the rest of the day it was smothered in cold clouds. The second morning I saw a curious effect. While the great mountain was still in cold blue shade, the rosy light coloured the clouds above it, and made them glow with fire; then the clouds in the valleys between myself and the mountains caught and reflected the colour from those upper clouds, carrying it down into the world below.

I took possession of my paintings at Jonboo the next day, and saw a most curious reflection of myself and the sun's disc in the mist, opposite the setting sun, with a gold halo and rainbow tints round it. It would have made a good suggestion for a Madonna or saint's picture. Darjeeling was still crowded, and Mr. and Mrs. D. squeezed themselves to take me in, most kindly, for a night; then I went down 2000 feet lower on the northeast side of the hill to Mr. H., the manager of several tea-estates. His bungalow was delightful: roomy and full of luxury, surrounded by the most lovely flowers, with gorgeous blue pheasants walking about a large aviary. Nothing could be seen for miles but tea-terraces and a little cinchona; the beautiful forest had disappeared to make way for it. I rode back to Darjeeling and then returned to Calcutta, where I was again picked up and taken possession of by a stray servant at the station, who fed me and locked up my door when I went out. I went off to the Botanic Gardens at daybreak, and got all I wanted at last. It is a very lovely garden, and I had more leisure to enjoy it than I had before. One great banyan tree was perhaps its greatest pride.

It was delightful after a long night and day in the dusty train to find myself again opposite Benares; and the weather had become perfect, neither too hot nor too cold. I hired a carriage for the week, and went every morning to sketch on the river, where there was such a mass of picturesqueness that the attempt to paint and reproduce it almost drove me to despair. Sitting in a boat anchored by a rope or a stone, with a fidgety man holding an umbrella between me and the cloudless morning sun, is not a comfortable mode of painting. After two days' work I began to hope the first sketch might give an idea of the thing, though a very superficial idea; and I began another with the Nepaulese pagoda in it, and some of the sliding temples, which had come down the hill quicker than they intended.

One morning I landed at the steep stairs leading up to the Nepaulese pagoda – a building which looked like a bit of Japan, and had a bell and cover of pure gold on its top. It was built on the top of a high platform guarded by towers and turrets. A huge tamarind and some peepul trees were planted round it, making a fine subject when seen from below, with boats of all sorts of odd shapes on the river in the foreground, the water being strewn with floating flowers, chiefly yellow ones.

Tomb of Ala ud Dïn and a Neem Tree (*Melia azadirachta*), Delhi, India.

A Vision of Eden

I left Benares on the 2nd of November, and passed through Khanpur at midnight, when I woke up to the delights of a cup of good coffee, poked in at the window, together with knives (open), toys, and pith models of the Memorial, reminding one of that awful mutiny-time. It made me wonder if it would not some day come again. No one can tell; we really know nothing of what natives think, and few make real friends among them.

On reaching Mathura, I found the one garry had been taken by the one Englishman in the train. I sat on the weighing-machine and waited amongst all the howling rabble till it returned. Such a lot of savage-looking faces. I fancied I would as soon settle at Mathura as any place I had seen in India for subjects for painting. It was so full of strange sights and curious groups of people and creatures. We went at five o'clock down to the Ghauts to see the turtles fed. They came paddling slowly across the river in hundreds, stretching their long snake-like heads out of the water and scrambling over one another eagerly after the food. The still more eager and adventurous apes came down also and jumped on their shells, trying to snatch at the food before the turtles had time to get it, and occasionally got their fingers badly bitten in their thievish attempts, for a turtle's bite is no joke. They even bite the natives occasionally when bathing; but the crocodiles did not seem to care for Hindus, and I heard of none being eaten by them.

At last I got to Delhi, which I had passed and resisted twice before, and Colonel and Mrs. D. kindly took me into their beautiful house for a week. The Royal City was a grand place, and quite came up to all I had heard of it. The first afternoon when I was driven round Delhi I saw the grand simple form of the huge mosque against a deep orange sunset sky, also the exquisite marble halls within the fort, with its massive red sandstone walls and gates. Also, alas! the hideous barrack buildings and other atrocities introduced by my countrymen. I worked regularly in the famous old marble palace which used originally to hold the peacock throne.

1878. – After a week here, I went on eleven miles, and settled myself in the bungalow which had been made in one of the old gate-towers, under the shadow of the Kutab, the highest tower in the world (250 feet), which reminded me of the American sequoias or big trees in its general proportions and colour. It has the same swell at the base, the same gradual tapering to the top, and I think it is one of the few buildings which really looks its full size. But all its surroundings are mere dwarfs, except the great pointed arches, built with horizontal stones overlapping one another, and no keystone. Near them was a forest of small Hindu pillars collected from six different temples, each pair of pillars of a different pattern. Many of the pillars are carved like cameos, but the conquerors decapitated most of the figures. The Gate of Ala ud Dīn was the gem of all; the details and lace-work of stone and marble, and the beauty of general design, surpassing any building I had yet seen in India. Beyond the actual Kutab ruins is a completely ruined city, with tanks and tombs and mosques without number scattered over some miles of ground, now left to the mercies of vultures and eagles. I had the place to myself for a week, except for chance visitors who came over for an hour or so, mounted the great tower, made a picnic meal, and drove away again; but after the first week the Davies arrived with their camp. Such a tribe of people, tents, camels, and horses! The Colonel had a regular court there, and all the neighbours (including beggars) turned out to meet him and camp round him. At night the place was lighted up with their fires.

After three nights the D.s rode away, their camp disappeared, and I remained in sole possession of the dead city and enjoyed its quiet. But the mornings got almost too cold for me to hold my brush, and I wandered about the ruins instead of attempting to sit and work, till the sun warmed the day. I found endless beauty in the old ruined city of Siri behind the Kutab. Some of the mosques were most elegant, with marbles and coloured tiles let into their walls and ceilings. In one building were exquisite medallions in fine plaster-work, which I longed to steal; nothing could be more beautiful than their designs, and they were only crumbling away there. Nearer me, the camp-horses had left an abundant harvest for the squirrels to gather in. They came in crowds to the place, and were most tame and pretty to watch. I nearly caught one little fellow, head downwards on a branch of a tree, nibbling at one of its pods as it hung. The cold also brought the mina-birds and doves down on the ground with their feathers all ruffled up, making believe they had fur coats on. It was sometimes rather lonely and awesome among those tombs of the old city, with the wild dogs, vultures, and bones.

On the 24th I drove away from the Kutab and across to Nizamudin, – a most lovely little tomb and mosque of white marble, with richly coloured praying-carpets hung up as blinds, and a grand old tamarind-tree over it. A fine old Moslem was keeping a school of little boys in the court, all the children gaily dressed, and standing on bright carpets spread on the marble pavements of the court as foregrounds. What a subject it was, if I could but have done it as I wished to do it! Another picture close by attracted me – a deep tank with tumble-down old buildings round it, and another old tree on the top of the steps, the faithful doing their prayers and ablutions on its edge. It was difficult to tear myself away, but at last I did, drove on to Delhi, and put up at a hotel made out of a collection of old tombs strung together, my bed being under a beautiful dome. I called at the post-office, and wanted to pay for a parcel of paints from Bombay – twelve annas. I had none, and offered two six-anna stamps bought at the same office, and was told by a native clerk "it was not a legal tender"! I felt inclined to box his ears. The priggishness of the educative native is most odious and ridiculous. The next morning after getting my money from the bank, I went on by rail to Alwar.

All the faces and costumes had completely changed in a few hours, and I was among a new race. The Rajputs were fine-looking men, with very marked features, long whiskers and moustachios, the ends of which they tied at the backs of their heads. They delighted in dyeing them red or orange too, which gave them a very odd look. They wore cashmere shawls on their shoulders and gay caps. Major L. met me at the station. He was most genuinely hospitable and good-natured (though peppery). His wife was in Canada with her own family. He took me the next morning to see the old Palace and its lovely tank before breakfast. The buildings were all of marble, with coloured lines and scrolls of mosaic work let into it. Peacocks were the wild game of the country; I saw hundreds on every road, eating up all the newly-sown corn, but they were never allowed to

be touched. The holy tank was full of ducks, geese, and swans. The latter looked most dignified, sailing under the red sandstone arches. Oranges and other fruits grew in abundance in the gardens, wherever there was sufficient irrigation. All the water was brought from a lake ten miles off, artificially made in the hills.

I saw a whole street full of hunting chetahs and lynxes. The latter are the most restless wild little animals, which jump at their victims' throats like bull-dogs and strangle them. I saw a man put a piece of meat on the end of a bamboo cane and hold it up as high as he could, and one of the small animals (not much bigger than a cat) made a prodigious spring from the ground to the top of it, took the meat, and was down again in an instant. Their ears stick up right over their heads and meet at the top. The chetahs are taken out in carts blindfolded and let out when within sight of the deer, when they creep up and spring upon them, holding them till the hunter comes up and kills them. They are so intent on their work that they are easily blindfolded and led away again. All those wild beasts are chained to trestle-beds in front of the houses down the street, their keepers sitting or sleeping behind them, and little children, peacocks, cocks and hens, wandering among them without the slightest fear.

Alwar was full of strange sights. I used to take my sketching stool and sit at the gate of an evening in the road, making studies of camels and passing travellers.

After a week the big-voiced Major drove me back to the station, and started me for Bhartpur, telegraphing to the Maharajah there to look after me and send to meet me, which he didn't; so I found my way to the bungalow in a country "ekkah" with all my luggage, my feet hanging over the wheels, for I could never learn to sit on my heels like "other people." When I heard the next morning that a large open carriage and horses had arrived in the night for me, I took for granted it was the carriage sent over from Mathura and I ordered it to take me to Dig (twenty miles); then I found out that the carriage I was in was not from Mathura, but was sent by the Maharajah. It had followed me to the bungalow, and was at my orders as long as I liked to keep it. The coachman did not appear surprised, though his horses were valuable ones.

One of the chief aims of Marianne North's visit to India was to build up a
collection of paintings of plants sacred in the literature and religion of India.
These are displayed in a group at Kew.

Opposite above left: The Indian Coral Tree (*Erythrina variegata*) is often mentioned by Indian poets
since Krishna stole it from the Celestial Garden for his wives; it has been under a curse, and is
never used in Hindu worship.

Opposite above right: The Frangipani (*Plumeria rubra*) is planted in cemeteries where the daily fall
of white flowers covers the graves with a fragrant carpet.

Opposite below left: The flowers of the Night Jessamine or Jasmine (*Nyctanthes arbor-tristis*) open at
sunset and fall about sunrise, and are much used to adorn temples.

Opposite below right: The fruit of the Mango (*Mangifera indica*) is widely eaten in tropical countries
and its wood is used by Hindus to burn their dead.

Above left: The Asoka (*Saraca indica*) is widely cultivated in India for the beauty of its foliage and
flowers. This piece, seen growing through a honeycomb, was picked by a priest of Karlee.

Above right: The juice of the leaves of the white-flowered Thorn Apple (*Datura metel*) was once
used by Rajpoot mothers who smeared their breasts in the belief that it destroyed their new-born
female children.

A Vision of Eden

When I reached that wonderful group of palaces, the guardian was asleep, and they could not find the keys. I was taken to a huge hall which looked very comfortless, all windows and marble; so I wandered downstairs to look at more rooms in another building full of exquisite carving, and suddenly found a big yellow snake in one of them, perfectly motionless, with his head up, looking at me! I was advised to keep in the upper rooms, as these elegant reptiles object to stairs and don't climb up them. Soon the keys were found, and I had the choice of any number of lovely rooms, with marble lacework for windows, and the finest carving on walls and ceilings. After a time I was told the Seth's carriage and four horses had also arrived in search of me from Mathura.

So I sent the Maharajah's back. The new coachman attached himself to me and made himself useful, giving me hopes of getting some dinner after sunset, when some hundred cups of oil were lighted and placed in every doorway of the passages and hall, as well as in a dozen or two of rooms leading to mine, and tall silver candlesticks were lighted in the hall which had the lovely windows. But at first, when I came in hungry, I found that rats had eaten my few remaining biscuits and left only some crumbs scattered about the table. I thought I should have preferred a dish of porridge to two hundred spare rooms. Soon after, however, gorgeous servants appeared, carrying in seven silver dishes full of capital food, with champagne, claret, coffee, etc. It was like a scene in the *Arabian Nights*, – even the poor fever-stricken guardian looked happy and gorgeous in a gold and silver livery. I stayed two nights there in the greatest magnificence, but was ashamed to stay longer. If I had had a quiet bungalow and bread and cheese I would willingly have stayed a month, the place was so beautiful.

Mathura seemed even odder and fresher on my second visit than it did at first, and I went every morning down to the ghat or street, among all the odd people, animals, and buildings, in my friend's carriage, working in peace and no hurry. The "Seth" of Mathura was said to be the richest man in all India. He had a weakness for keeping beautiful carriages and horses and lending them to Europeans, and he had sent over to Dig for me. He sent me back straight to Bhartpur, twenty-three miles, with three pairs of splendid horses to

relieve one another at different points in the road. From thence the rail took me to Jaipur, where Colonel B. received me most kindly at the Residency – a beautiful oriental house in a garden full of cypress-trees and fine shrubs. The Colonel was a model English gentleman and a great friend of the odd old Maharajah, who came to see him the afternoon after I arrived and made me bring in my Mathura work to show him, though he was nearly blind. I liked the odd old man, and he seemed thoroughly to like the Colonel, sitting on the sofa beside him with his hand on his shoulder half the time. He promised to send his biggest elephant in all its most gorgeous attire and ornaments for me to paint the next day.

The gardens of Jaipur are lovely, with the stony hills in the background. A drive of seven miles through picturesque suburbs and between gardens took me to the foot of the hills, where I had the option of a tonga with bearers or an elephant to take me up the steep road, and on by the side of the lake to Amher, the old capital of the country. It is all now deserted, but crammed with subjects for painting, – fort, palace, temples, tanks, and gates, among bare stony mountains, with distant views of the sandy plain. Near the lake were great india-rubber and banyan trees, with their roots twisted about in all sorts of unexpected places and hanging in long fringes from their branches. It is a grand place, and I could not understand why they ever left it for a low place like Jaipur, where a perpetual war is going on with the encroaching sand-drifts. Jaipur itself is the second largest city of India. I spent one afternoon painting the state elephant in the garden. He was covered with crimson and purple velvet heavy with silver embroidery and kincob. Two golden lions were sitting on the top of his head, his face was covered with silver plates and bells, while on his ears dancing-girls were painted. I wanted to paint him also kneeling down, but he moaned so much over the cramped position that I let him off. All the time he did kneel, his attendants held the gorgeous draperies out like a tent so that they should not touch the ground (four men's work, they were so heavy); the whole was only a few inches off the ground when he stood up, which saved me the difficulty of drawing his legs.

The Brahmin bulls were a great curse in Jaipur. They used to go up to the poor market-people's

India

The Bridge of Chitor, Rajpootana with a tomb and Tree of Heaven (*Ailanthus altissima*) in the foreground.

baskets and eat up everything without even saying "by your leave." They were noble-looking beasts, and everybody (who was anybody) went into mourning when one died. His heir was given, with a cow, to the ruling Brahmin of the place to keep; but bulls do not give milk, so the holy man only kept the cow, turning the bull loose into the streets to support himself as best he could by stealing from the devout Hindus around. In the famine time it was quite scandalous the amount that was consumed by the idle beasts. Colonel B. persuaded the Maharajah to lock them up and put them on scant rations, thereby causing a small rebellion among the people of the place at the insult offered to the souls of their ancestors. They swore that famine was caused by the way the sacred bulls were treated, so they were let out again.

When I reached Ajmere I heard at the bungalow that Captain L. had called an hour before, so I drove out to Mayo College, where I found him and his bride most hospitable, and they made me come that same evening to stay. Every other European was in camp. Captain L. had been one of Lord Lytton's A.D.C.'s. He had just married and settled as Head of the New College for young Rajahs, and his honeymoon was barely

over. The place was on a sandy plain, surrounded by bare hills, whose tints were glorious at sunset. It was two or three miles from old Ajmere and its lake, but the college interested me too in its way, as an experiment of young Indian education.

Captain L. got the native authorities at Ajmere to write on my behalf to the Governor of Chitor for a permit and protection when sketching in that wonderful old ruined city. He drove me over to Nasirabad himself, and arranged with the magistrate to send me a chaprasi and to get me a camel-carriage; then after a cup of tea together in the bungalow, he and his wife drove back and left me to my fate; and at night I started, those old beasts, the camels, preferring to tramp all night and munch all day.

The first day I spent in a nice airy bungalow, on a rich plain. Cotton was the commonest crop of the country. Women in long blue veils and dresses edged with red were picking it, hidden among its stalks, for it grew as high as themselves. Their arms were nearly covered all the way up with bangles, chiefly made of lacquer. The cornfields were also looking green, which was refreshing after the long desert we had come over, with nothing but euphorbias, and now and then a barbal or acacia-bush, to enliven them.

The second night's journey brought us to Bhilwara, to find the bungalow surrounded by a

145

camp of English artillery, with guns, camels, and even elephants. The chief officer was sitting on the verandah and occupied the room next mine. I spent the day drawing their troop of camels being loaded, and saw all the tents packed on their backs, and the soldiers' wives with clean white aprons, arms akimbo, and helmets like their husbands to preserve their complexions, cooking their Christmas dinners, while the jolly English soldiers nursed the babies.

Another night's tramp and I saw the long hill of Chitor before me, arriving about ten o'clock at the bungalow on the plain beneath it. Before I had time to get my hat off, a most genial English gentleman, Colonel C., was at my door, with his hands full of English letters and papers for me. He said he and his camp were close by, and he had expected me to spend a merry Christmas with him the day before. He was to leave the next day, but before he left he made all sorts of kind arrangements for my stay and future movements, and I had breakfast with a merry party of friends whom the Colonel had visited to spend Christmas with him.

The whole hill-top was surrounded by strong walls, towers, and precipices. At the foot of the steep ascent was a considerable Indian town, and beyond it a broad river, crossed by a noble bridge of pointed arches, approached at either end by a steep incline paved with great stones. The rest of the bridge was quite flat, and paved in the same way, with no parapet. It was so slippery that natives and animals preferred the safer course of wading through the water. The ferry beside the bridge was always crowded with splendid subjects for a figure-artist, the whole making an exquisite picture, with palms and different rich foliage on the banks, the hill behind covered with a kind of scrub, quite purple in its autumn dress, and looking even crimson at sunset: above all the old ruins and towers of Chitor, while great squared stones were strewed about the river's edge and bed.

While I was there the Maharajah arrived with all his train on a hunting expedition. He passed the lake one day where I was painting, with all his hunting procession; but I hid myself behind the rocks, as such grandees have a habit of demanding anything they have a fancy for, and have no idea of being refused, while I had no idea of giving away

the sketch I had come so far to make, to a half-civilised human being I had never seen before.

That night I was awakened at one o'clock in the morning by a great thumping at the door and calls for "Mem sahib." I had to get up in the cold, out of my warm quilt, and open the door, to see two tall white bundles and a still taller camel, looking all dislocated, as camels do when tired. A letter was thrust into my hands, and I did not bless them, but shivered and grumbled ungratefully till morning. Those natives never think, they only do what they are told; and of course they thought Colonel C.'s despatch of the utmost importance. In it he had written my exact programme of proceedings for the next day; and when such arrangements are made by great men in India it is wise to keep exactly to them, or one gets into trouble and difficulties. I had one more day on the old hill, and saw his Highness go off on a tiger-hunt, with a long procession of elephants and led horses. As I passed the chief temple the steps were running down with blood from some great sacrifice in his honour, he himself having killed the poor beast with his royal hand. The great feat was to strike off the head with one blow of the sword.

At last Dr. S. arrived with the Maharajah's big open carriage and four, and we started together at a grand pace. At the last stage an elephant was sent out to meet him by the greatest noble of Udaipur, whose brother he had been to doctor; but Dr. S. preferred the easy carriage to the slow lounging pace of the more dignified elephant, and we passed through the great gate which fills up a chasm in the hills seven miles from Udaipur (and is the only entrance on that side) just as the last remnant of daylight faded from the sky. The rest of the drive was made by the light of the stars. Even in that light I could distinguish the city glittering like a group of pearls, with the marble palace above it, and the lake behind, surrounded by bare mountains. Then we turned up a side-road to the Residency – a picturesque mass of Oriental buildings, fitted with English comforts, which I had all to myself, as the Doctor went to his own quarters, in a tent, half a mile off.

The next morning he came early and took me in the grand carriage, which was put at my disposal

Ginger Lily (*Hedychium gardnerianum*) and Firetailed Sunbird (*Aethopyga ignicauda*), India.

India

Above: A ruined mosque at Chāmpanir, near Baroda, Western India.

Opposite: Foliage, flowers and fruit of the Pride of India (*Lagerstroemia speciosa*), one of the most magnificent forest trees of India.

all the time I stayed, by the Rana, and we drove through the old gates into the city and through the busy bazaars to the grand Jaganāth Temple in its centre – one of the richest bits of Hindu work I had seen in India, great elephants of stone guarding the top of the steep flight of steps which lead to it. Then we walked down a steep street to the water-gate, and the full glory of the lovely lake burst upon us, with its distant islands of palm-trees and marble palaces, and its nearer orangery surrounded by white marble arches and pavilions with exquisite tracery. Still nearer, palaces, gardens, and gates, all reflected in the still blue waters, and over all the pale salmon-coloured hills, with their lilac shadows, so faint, yet so pure in colour. We found the Rana's boat waiting for us, rowed round all the islands, and walked in their lovely gardens and courts. In one of these sixty English women and children had been sheltered, clothed, and fed during the mutiny by the good Rao of Baidah. He went with his drawn sword to the Rana, and said he would kill him if he did not help them; then took an escort and rode off ninety miles to Nimach, and brought them safely back with him. It was a small paradise, but no doubt they thought it a prison. This good old man had a great liking for Dr. S. He came and paid me two visits, looked at my paintings (and held none of them upside down). He was a most noble old man, full of intelligence, and he told me many things I did not know about the different plants whose portraits I showed him. We landed on the mainland, and mounted by steep terraces and steps to the very top of the huge marble palace. No views could be more superb than those from the upper storeys, over lake, city, river, and mountain.

The great Moslem fête of the Muharram took place while I was at Udaipur. The Doctor came and fetched me from the Palace (where I was at work), and took me to a room over a gate in the High Street, where the chief banker, his boys and friends, were looking out at the crowd. I made a sketch of the whole procession winding up the narrow High Street, with the Palace and great Temple against the sky above it.

I found that twelve hours of work daily in boat and on land, without coming home to rest, was too much even for my strength. I had to take a day in the house; when the Rao came again and paid me a long visit, saying he hoped to meet me again, if not in this world, in the next; and the Doctor said that if he were not worthy of a good place there, no Christian was.

At last I left the wonderful city, so full of magnificence, and yet not three hundred years old! For sixty-four miles I was carried on in a great open carriage, full gallop, with four fresh horses awaiting me every twelve miles. After that the chaprasi Col. C. had sent me, took his departure joyfully to rejoin his master, being fairly tired of the sketching woman.

I returned to Bombay, where Mrs. R.C. gave me a most kind welcome and put me in the nicest of all the bungalows at Government House, on the furthest point of the cliff, with sea on three sides of me. After a few days' rest I took my passage home, then started by rail three hundred and nine miles northward to Ahmedabad. I had only twenty days more in India, and I wanted to make the most of them.

My next move was back to Baroda, and then to Chāmpānir. My journey back to Baroda was not

without its dangers. Once we stopped, and I saw my guardian and the two mounted police and four odd running men with torches all drawn up in a row in front of my bullocks, which shied right round as two small donkeys trotted out of the jungle! And the drawn swords, bows, and arrows were put out of the way again, and my last chance of seeing a tiger was gone too.

Before I left Baroda I received a letter from Major Nutt, saying he expected me to cross by the next steamer from Surat to Bhaunagar, where I should find the carriage of the Maharajah to take me over the eighteen miles to his house on my way to Pālitana. This offer I took up before returning to Government House at Bombay for two days' shopping and packing.

On the 24th of February 1879 I went on board my old friend the P. and O. ship *Pekin* again, and, thanks to the kindness of its officers, got a good cabin to myself, and had a calm voyage to Aden, where the colours seemed more lovely even than my remembrance of them, and everything about it attracted me as before. Cold was the one enemy I dreaded, and I met it in the steam-launch which took us up Southampton Water. What an icy wind it was! I gave my keys to some man on board who said he was an agent, just caught the train, and reached home with my little bag in my hand on the 21st of March 1879, leaving luggage to follow the next day. It was very luxurious being again at home and amongst all my kind friends, but I found the perpetual task of showing my Indian sketches very wearisome. Every friend brought or sent other friends, and so many were interested in India that my tongue got no rest from telling the same old stories.

At last I hired a room in Conduit Street for July and August, and General M'Murdo most kindly helped me to start an exhibition there. This paid two-thirds of the expenses by the shillings taken at the door, and the remaining third I thought well spent in the saving of fatigue and boredom at home.

CHAPTER IX

Second Visit to Borneo – Australia

1880–81

AMONG THE CRITICISMS of my paintings in Conduit Street was one in the *Pall Mall Gazette*, which suggested that the collection of botanical subjects should find their ultimate home at Kew. I kept this idea some time in my head before acting on it; but having missed a train at Shrewsbury one day and having some hours to spare, I wrote off to Sir Joseph Hooker and asked him if he would like me to give them to Kew Gardens, and to build a gallery to put them in, with a guardian's house. I wished to combine this gallery with a rest-house and a place where refreshments could be had – tea, coffee, etc.

Sir Joseph at once accepted the first part of my offer, but said it would be impossible to supply refreshments to so many (77,000 people all at once possibly on a Bank Holiday), mentioning, too, the difficulty of keeping the British Public in order. I asked Mr. Fergusson, the author of the *History of Architecture*, to make the design and manage the building for me, which he did to the end with the greatest kindness and carefulness. I chose the site myself, far off from the usual entrance gates, as I thought a resting-place and shelter from rain and sun were more needed there, by those who cared sufficiently for plants to have made their way through all the houses. Those persons who merely cared for promenading would probably never get beyond the palm-house. There was a gate and lodge close to my site for those who drove there straight, and though that gate was kept shut then, I hoped to get it opened by means of the *vox populi* in due time – perhaps not in my lifetime. I also obtained leave to build a small studio for myself or

any other artist to paint flowers in at any time, as there was no quiet room in the gardens in which a specimen could be copied, away from the sloppy green-houses and traffic of visitors.

One day, after arranging all this, I was asked by Mrs. Lichfield to come and meet her father, Charles Darwin, who wanted to see me, but could not climb my stairs. He was, in my eyes, the greatest man living, the most truthful, as well as the most unselfish and modest, always trying to give others rather than himself the credit of his own great thoughts and work. He seemed to have the power of bringing out other people's best points by mere contact with his own superiority. I was much flattered at his wishing to see me, and when he said he thought I ought not to attempt any representation of the vegetation of the world until I had seen and painted the Australian, which was so unlike that of any other country, I determined to take it as a royal command and to go at once. Mrs. Brooke persuaded me to return with her and the Rajah to Sarawak, and make a half-way rest there; so I joined her party on board the *Sindh* at Marseilles on the 18th of April 1880, and arrived at Singapore on the 15th of May, after an agreeable voyage in that most excellent French steamer in which I had once returned from Ceylon.

We stayed near the beautiful Botanical Gardens, and used to take a stroll there in the cool evening, or sit and hear the band and scandalise our neighbours in the orthodox manner of the place. The ground-orchids were magnificent in that garden, with flower-stalks as high as myself. There are also a good many animals and birds;

amongst the former a huge ape, who put his long arm through the bars of his cage without any warning and grabbed hold of anything he fancied with irresistible force. He only got the button of my umbrella from me, but had quite lately seized the watch-chain and locket of a German gentleman, dipped them in the water, and then munched them slowly, while the German danced round and round like a madman, lamenting the portraits of his beloved ones on the other side of the world, helpless to save them, till a native seized one of the hanging ants' nests from a tree and flung it at the brute, which dropped the mangled treasures with a savage growl as the small creatures revenged on his body the injury done to their house. Those nests were as big as two heads, made of leaves sewn together most cleverly. One hung close to our bungalow on an alamanda bush, and made a pretty picture, surrounded with its bunches of lovely yellow bells.

However, because of the heat and mosquitoes, we were not sorry to escape in the Rajah's gunboat, the *Alarm*, to Sarawak, where we arrived on the 25th of May. To me the Astana was even more attractive than before; for the plants in the garden were older and larger. Some of the views might be lost by that, but the whole effect was even more luxuriant. It was still surrounded by its rim of impenetrable forest-tangle, and the great trees made the most harmonious background to the gorgeous shrubs and creepers, on all but one side – that which bordered the ever-moving river, covered with its busy semi-amphibious people, in every variety of canoe or boat. The Rajah was as happy to be at home again as his people were to see him after his two years in England. His dominion got more prosperous every year. The last census was about 200,000, the income £45,000, the country about the size of Scotland; and this strange mixture of races submitted cheerfully to the mild despotism of one simple Englishman. In the quiet mornings I found delightful studies close by. My first was a boat-house, with trees and

Flowers and butterflies of Sarawak. The butterflies (*Ornithoptera priamus*) are on *Mussaenda macrophylla* with two species of *Clerodendron*, one with blood-red flowers and one in fruit.

palms half-buried in the water at high tide. Children were running in and out, regardless of crocodiles, with no clothing but their necklaces, pouring the water over their heads with monkey-cups (pitchers of the nepenthes). They were full of fun, and such lovely round shiny little mortals. Why don't real artists go to paint them?

I was taken one day for a drive by the Rajah over the fifteen miles of road he has made on the Kuching side of the river. At present the road leads nowhere, and goes through no place; but Chinamen are expected to follow it, and to build themselves villages, and it does to exercise the horses on (which are about as useful in Sarawak as they are in Venice, the high-roads being all water, with this one exception). We saw much fine forest, and I succeeded in getting at last a specimen of the clerodendron, with orange leaves or bracts, one of the most singular flowers I ever saw. I had seen yards of the orange leaves, but never a green one till then, and it unfortunately grew in such inaccessible places that I could not get at it myself.

Mr. B. again took me up the river in his steam-launch to Busen, and on with the tram and a pony to Tegora, through that wonderful forest, with its dangerous bridges. One broke in two just as my pony scrambled over, the strut under it having been washed away by a late swollen stream. I was on foot, so I came to no harm. There were no flowers, but the coloured leaf-chains and rattans clinging to the trees were most lovely. I spent a day in a canoe, painting the red-stemmed Palawen trees, which grew only on that river. The trunks were quite flame-coloured in the sun; the bark hung in tatters from them, and from the branches, as on the eucalyptus. The leaves were like those of the laurel, with brown young shoots, but I could get neither seeds nor flowers.

I began my return journey badly, for my pony displaced a plank in one of those horrid bridges, took fright, ran away, and tumbled me off, the Rani's new saddle having no off pommel, and I lost my spectacles in a bank of fern. My friends too were frightened to let me risk the other slippery bridges, and got a canoe with a mat and pillow, and two Dyaks to paddle and push; so I gained once more the pleasure of shooting the rapids, lying on my back and looking at the tangled branches overhead, with their wonderful para-sites; sometimes shooting swiftly down through

A Vision of Eden

Above: A Tapang or Tappan Tree (*Koompassia excelsa*) of Sarawak, whose trunks often rise 100ft without branching.

Opposite: The Turong Bird or Pigeon Orchid (*Dendrobium crumenatum*) of Sarawak comes into blossom simultaneously on all plants about every nine weeks but the snowy display lasts only one day. The purple-brown orchid is *Cymbidium finlaysonianum*.

deep-green water and white foam, while the men clutched at the rocks and tree-stumps – sometimes being almost carried by them over a few feet of water. At last we reached the place where Mr. B. was throwing in big stones to make a dry landing-place for me; after which came a pleasant ten miles of forest-ride and no accidents, and I was left to stay three days with the young manager of the antimony mines at Busen, where I spent my mornings up the river in a canoe, sketching strange trees and a bamboo bridge. We saw some grand specimens of the Tappan trees, with their smooth white stems, on which bees delight to build their nests. No beasts or reptiles can climb these trees, only the Dyaks beat the bees by building clever ladders with bits of bamboo. They drive in pegs above their heads, tying the other end of the peg to long slender bamboos, placed one over the other and lashed together with pieces of bark, till they reach the honey and wax. The latter being one of the great exports of the country, these trees are protected by the Government.

The steam launch came up for me and carried me far too fast down the beautiful broad river again. At one point we saw a water-snake full twelve feet long, with its head held nearly a foot out of the water, swimming across most gracefully. It was all red and green, with a sharp ridge down its back. My sailors wanted to steam across it, but it was too quick for us, and curved its beautiful head back to dart an indignant and contemptuous glance at us as we missed it, perfectly ignoring the odds and ends thrown at it. The old steersman told me it was very wicked and poisonous, but it looked all in character with the other surroundings of the great river. I found the Rani delighted to see me back, as her husband had taken her brother and his shadow off, up one of the other rivers, in the gunboat for a fortnight, and she was very lonely, but busy all day trying to paint as I did. Bushels of rare flowers were brought in, and died in two hours in the attempt, but she said life was twice as bright since I had set her trying to copy them, and it was a comfort to think I had done her some good by coming, for her life was monotonous, the brightest days being those which brought the mail with news of her boys in England.

There was a great *Vanda lowii* which had a

154

A Vision of Eden

dozen sprays on it at once, each eight feet long, the year before, I was told. I watched it from day to day so anxiously; and one morning (after the Rajah returned), to my horror, I found the whole orchid-house (a mere skeleton wooden erection) flat on the ground, and the great ground-orchids mown down also! The other lovely orchids and trailing plants, stephanotis, etc., in full flower, all fading or dead, by order of that "mild despot" the Rajah! I felt glad I was going, but as usual he was right. I heard that three months afterwards the mass of beautiful flowers was even more luxuriant than before.

One morning I picked a huge branch of the *Petraea*, meaning to spend the day in painting it, though it was so common there, when I came on a lovely spray of a white orchid and picked it grudgingly to paint, then suddenly found that every tree was loaded with the same, and the boathouse roof looked as if there had been a sudden snowstorm. The air was scented with it, so I got more, and when I reached the house found the drawing-room full of it. They called it the Turong Bird, and said it came out spontaneously into bloom three times in the year, and only lasted a day, and that I must be quick and draw it, for I should find none the next day. It was true; the next day the lovely flowers were hanging like rags.

When I went to finish another sketch I was astounded at the sight of a huge lily, with white face and pink stalks and backs, resting its heavy head on the ground. It grew from a single-stemmed plant, with grand curved leaves above the flower, and was called there the *brookiana* lily, but Kew magnates call it *Crinum augustum*; its head was two feet across, and I had to take a smaller specimen to paint, in order to get it into my half-sheet of paper life-size. It was scented like vanilla. Another crinum has since been called *northiana*, after myself. It has a magnificent flower, growing almost in the water, each plant becoming an island at high tide, with beautiful reflections under it, and its perfect white petals enriched by the bright pink stamens which hang over them.

On the 10th of July I left the beautiful island again, the Rajah putting me on board himself, like the true English gentleman he was. Two nights and a day brought us to Singapore, where I was received by the colour-sergeant and his wife.

On the 19th of July the *Normandy* steamed out of the harbour – the very smallest of comfortable ships, with a most agreeable and sociable small company of people on board. There were also two ponies, a puppy, three cats, and three monkeys to entertain us. That ship was a perpetual comedy. One day the big monkey was discovered sitting solemnly on the captain's arm-chair in his dark cabin, reading Shakespeare, after having rubbed his face and paws in the fresh paint outside, smeared over all the photographs and pictures on the walls, and overturned the clock. When scolded he did not move, but tried to lick off the paint from his hands in the same way a naughty boy might have done.

We were a week out of sight of land, and it was so cool the day we crossed the Equator that we were glad to wear woollen dresses. At last we came in sight of the little round island with its cave, in which provisions for possible wrecks used to be put, as well as letters for ships to pick up and take on. After dark we got slowly into the harbour of Thursday Island. Five pearl-captains came on board and went on with us, and one of them gave a basket of orchids to the captain. I tried in vain to paint one flower, it being too stormy on deck and too stifling in the cabin.

The morning we passed through the Torres Straits was calm and lovely. We were so close to each shore that we saw all the trees perfectly: mangroves, casuarinas, brown granite rocks and red sand, with giant ants' nests like obelisks of the same colour. At Somerset we saw the big house on the hill, with the ground all burnt bare in front to drive away, or at least make visible, the deaf adders, which are the curse of the place. We spent a day in the harbour of Cookstown, which is perfectly surrounded by wooded hills and islands, the town out of sight. If I had known of the place beforehand I would have waited a mail there, and got some paintings of tropical Australian vegetation, but no one knew anything beforehand.

On the 8th of August we reached Brisbane, passing the bar – a long treble line of breakers set in

View in the Brisbane Botanic Garden with a Moreton Bay Pine (*Araucaria cunninghamii*) in the foreground with in front an American *Passiflora*.

156

sapphires and emeralds – without difficulty : when rough the troughs between the surf nearly swallowed the ships, but that day all was easy. The cold was most bitter, and we were told there had been two inches of ice up the country. My fellow passengers packed me into a cab, waved their hats, and cheered as I drove off. Not a bad start on the Australian Island.

1880. – The Hotel was clean but unattractive, and over full. Next day I was not sorry to accept Mr. and Mrs. J.B.'s kind invitation to stay at Government House, whose garden opened into the Botanical Gardens. The old director, Mr. Hill, also offered me a room in his house, and put a paragraph about my paintings in the local paper, which induced a little boy to ask, "What will they do with you when you return to England, will the Queen knight you?" (That little Australian boy's question was answered four years later in a letter from the Queen's Private Secretary, in which he expressed "Her Majesty's regret at learning from Her Ministers that Her Majesty's government have no power of recommending to the Queen any mode of publicly recognising Miss North's generous gift to the nation" (by knighthood or otherwise). The Queen therefore sent her own photograph instead – a graceful gift, which gave her (ill then and depressed) the keenest pleasure.) The weather was too untropical for much out-of-door sketching, and the gardens were dried up and unattractive. Hot sun, cold wind, and dust. The famous araucaria trees in the Botanical Gardens were brown and dusty, and not larger than the one in the temperate house at Kew. The ferns and palms looked bare and cold. There were few flowers, though the Government House garden alone was rich with sweet home flowers – roses, carnations, heliotropes, etc., a few tecomas and tacsonias in addition showing that the present cold was rare. Dracaenas, strelitzias, and Norfolk Island pines also give a different look to the gardens, as well as wattle trees, yellow, with thousands of fairy balls and leaves mimicking the eucalyptus, though the young seedlings begin with the ordinary acacia leaves.

Brisbane itself is a most unattractive place – a sort of overgrown village, with wide empty streets full of driving dust and sand, surrounded by wretched suburbs of wooden huts scattered over steep bare hills. The hospitality was of the heavy

order, – great luncheon-parties, with soup and fish, and four corner-dishes, roast and boiled, etc., as it was forty years ago in England, – and I was glad to escape to the hills. Mr. and Mrs. B. got up at six to see me off, and had arranged that their cousin (who managed their property) should show me the trees I wanted most to see on it.

At Dalby Mr. and Mrs. M. met me with a troop of nice children and a pair of well-bred horses, and after a good English tea at a pretty little wooden many-roomed inn of one storey, hung with cages of parrots all round its verandah, we trotted over the plains towards the hills, picking up a quantity of parcels on our way at "the store." Mrs. M. and her daughters were all model housewives, and quite independent of all servants or outside help of any sort ; they waited themselves at the table. The farm-men ate with us, and everything was exquisitely clean and well arranged. Sam, the Chinese cook, and some black loafers, seemed their only servants. The blacks lived in their own tiny wigwams near, received no wages, and had no particular work, but made themselves generally useful, getting occasional presents and scraps in return. They were devoted to the children, liked looking after horses, and were entirely trusted by the family they had adopted, but did not think it wrong to steal from others. I was only allowed one day of rest, and then we started for the Bunya Mountains, leaving two fair young girls and the black people only on the premises. They said they were quite safe, but their companions looked a queer lot, with their shock-heads and short pipes. The tents and heavy things had gone on with a long team of horses the day before.

Mr. M. drove his wife, myself, and three children, in a kind of inside-car with a roof and four posts to hold it. More children and three gentlemen rode, and there was a spring-cart full of eatables and luggage which broke down in every gully. Those gullies were no joke to get down into, and harder still to get out of. No civilised driver in Europe would have attempted it, but Mr. M. never had an accident, and his black scouts moved frantically backwards and forwards,

Nest of the Coachman's Whip Bird (*Psophodes crepitans*) in a Bunya-bunya (*Araucaria bidwillii*).

158

finding the road, or chopping down trees to make one, or patching up the unfortunate spring-cart. In the afternoon we reached the steep ascent where we were to leave the carriage. I was mounted on the best horse, and started ahead with Sam the Chinaman, who had been disagreeing with his horse all the way, and was in an awful temper now. It was his usual state: no one minded "Sam," who made the best bread in all Australia.

I soon passed him, and much enjoyed my entire solitude through the grand forest alone, especially when I reached the magnificent old araucarias. Their trunks were perfectly round, with purple rings all the way up, showing where the branches had been once, straight as arrows up to the leafy tops, which were round like the top of an egg or dome, and often 200 feet above the ground. Only the ends of the branches had bunches of leaves on them, and only a third of the stem had branches left on it. But these grand green domes covered one hundred miles of hill-tops, and towered over all the other trees of those forests. Nowhere else were the old bunya trees to be seen at all; and at the season when the cones ripened, the native population collected from all parts and lived on the nuts, which were as large as chestnuts. Every tree was said to belong to some particular family, and they produced so great an abundance of fruit that it was also said, the owners let them out to other tribes on condition that they did not touch the lizards, snakes, and 'possums – a queer form of game-preserving, which reduced the hirer to such a state of longing for animal food that babies disappeared, and then there was a row, and no white person ever ventured on those hills while the bunya harvest was going on. Under these giants there was a fine undergrowth of every shade of green, brown, and yellow, roped together by fantastically twisted lianes. After half an hour of this scrub I came out to the clearing, where our tents were being put up, having arrived only a little before us, owing to the difficulties of the road.

I had my first sight of a party of perhaps twenty kangaroos, all hopping down the hill in single file, or feeding in the hollow below. I can fancy no more comical sight than a procession of these strange creatures, proceeding over the long tufted grass in the way I saw them then, using their big tails for balancing-rods. Another day we rode

farther into the forest, and saw still bigger bunya trees, and great skeleton fig-trees hugging some other victim-tree to death, with its roots spreading over the ground at its base like the tentacles of some horrid sea-monster. Great piles of sawdust and chip, with some huge logs, told that the work of destruction had begun, and civilised men would soon drive out not only the aborigines but their food and shelter. Under the trees were many of the leafy mounds made by the brush-turkeys to put their eggs in, and in which these are hatched by the heat produced by natural fermentation, without the trouble of sitting on them. The flesh of the bird is brown, and has a game flavour.

A poor little sloth-bear, was shot for me before I could say "don't" – so soft and harmless, all wool and no body or bones. I felt so sorry for the useless murder. They also burned the grass, and the fire came alarmingly near the tents; but the trees are so full of moisture that they never catch fire till after many days of scorching, while the grass blazes up and is out at once. In this case care was taken to pull up a circle round us before the match was thrown in. When by accident the flames come too near, every white man, woman and child has to take branches and beat it out, while the blacks sit down and sigh. The young grass is stifled by the dense mass of dry tufts above it. The only way of giving it necessary room and air is by burning off the old grass, and its ashes are the best manure for the young shoots. On our way down I again started alone on foot, stopping to enjoy and examine all the lovely things in the forest, and to make pencil-studies of leaves and plants.

The xanthorrhoea or grass-trees form perfect globes, and it was most delightful to pull the young centre spikes apart and let the thousands of fine green hairs free. The peeps through them on the blue world below were enchanting. When I got there I spread my pocket-handkerchief on a tuft of grass for a pillow, and lay on my back examining the Eucalyptus leaves overhead for an hour at least before the others came, in perfect peace. It is a libel to call them shadeless trees. Just at noon the knife-edges of the leaves are turned towards the sun, thinking more of keeping themselves cool by exposing the least possible surface to the sun's rays, than of shading the ground below. But in the morning and evening they give more shade to their own roots than

Second Visit to Borneo – Australia

European trees, and their constant movements fan the air pleasantly: the scent is delicious.

It was pleasant resting at the M.s' farm again. They were an ideal family for bush life. But the pet of all was "Jo," a young lady of eight years old. She actually offered to climb to the top of a gum-tree and bring me down one of the tiny native bears alive, in spite of scratches and bites! The whole family had a wondrous power of eating jam. Three pots were emptied every morning at breakfast. There are few cases known of delicate appetite in the bush, and the amount of meat consumed was appalling.

The return drive across Darling Downs was bitterly cold. The railway took me on to Harlaxton, a house on the very top of the zigzag railway, with a glorious view over "the Range," as the hills are called between it and Brisbane. My next move was westward by rail to Gumbara, where Mr. W. met and drove me from the station over eleven dreary miles of plain to his luxurious little house and pretty wife. I was told if I heard noises on the roof over my head at night I was not to mind them; it would only be a snake hunting an 'possum which lived there. How did the snakes get there? The laughing jackasses (a giant kingfisher) pounced down on the snakes whenever they had a chance, but often dropped them on the roofs when they could carry them no farther, and then the 'possum suffered.

My next move was to Warwink. Here Mr. G. met me and drove me on to sleep at an old house where there were fifteen daughters and nieces assembled, and I began to think that I had misunderstood the accounts of the country, and that women, not men, preponderated there. It was a thirty miles' drive on through the bush to Maryland, mostly without a road; but I was told that buggies could not turn over, and was taken in and out between the trees and up and down steep ditches and banks in the cleverest way. A noble old "carragong" tree stood near the house at Maryland. It had a rough reddish bark, with bright-green ivy-shaped leaves, which made it very conspicuous among the gray gum leaves, with their smooth marble-like trunks. I saw also fine casuarina trees called oaks, but looking like feathery cedars. They have a peculiar mistletoe which mimics them.

I was taken to Starthorpe to see an aviary full of parrots in an old lady's garden there. She had a dozen different sorts, some guinea-pigs, quails, doves, and a New Guinea cockatoo with red points to its white plumage – all in one big cage, and all living in perfect harmony. Starthorpe was a long straggling street of the usual one-storeyed detached shops, and seemed a busy place, owing to the tin-mines all round. A few miles farther took us to the principal one. We waited at the bridge till Cobbe and Co.'s coach came, with Miss B. keeping my place on the box, where we were shaken and pounded till after dark. We passed much grand forest-scenery, and one romantic bit between walls of huge granite boulders called "the Gap," was full of ferns. We saw three dead black snakes, and one live one, rearing its ugly head and neck out of a low bush to watch us as we stopped to look at it. The driver wanted to get down and kill it, but said the horses would be mad if they saw it. He said he would only give it a slight knock on the head with a stick, and it would then bite itself and die of its own poison. The blacks, when they want to eat one, take good care to kill it off at once, not giving it time to do that, as the flesh would also become poisonous – an odd story, if true!

At Tenterfield I had the usual fuss when two persons travel together, and was told only one room was to be had, but I was obstinate, and ultimately got another. It was what Australians call "a very pretty place," meaning that there was not a tree within a mile of it, and that it had a little water within reach. We were said to be the first ladies who ever travelled on that road alone by Cobbe and Co.'s coaches, so Cobbe and Co. were proud of us, and telegraphed beforehand to all the halting-places to have an extra quantity of beef ready for us, as well as that horrid double-bedded room. Cobbe and Co. coaches all Australia, with extraordinary vehicles of every shape and size, and really splendid horses: I hardly saw a bad one belonging to the famous company. The next day's journey took twelve hours, and I was so stiff that I fell flat down in the street when I attempted to get off the box. But the day had been most enjoyable, in spite of the squeezing and jolting. Soon after sunrise we overtook a tiny native bear on the road, sitting calmly within a yard of the wheels. We stopped the coach, and the driver made him a speech beginning "little man," which the small woollen ball thought impertinent; for after staring a few minutes, he suddenly got on his four feet and scrambled off towards the trees, faster than I thought possible on his funny clown-like feet. He had huge furry ears, and a most fascinating expression of face. The dryness was sad to see. But on this third day's journey in the coach the rain began, and though it made the roads heavy, we all rejoiced for the sake of the dried-up thirsty country.

Armadale we reached after dark – a considerable place, with some stone houses in it, and a bishop. An English clergyman met me at the door as we entered the hotel, with a note from Mrs. Marsh. The hotels were all good, and the charge, ten shillings a day, included a sitting-room and good fire. Meat cost a shilling for ten pounds. We started at eight, outside a large coach, with four noble black horses to pull us over an excellent road, as straight as a piece of Roman work, to Uralla, where we descended from our high perch, had a good breakfast, and waited till Mrs. M. came for us. We had passed some acres of orchards, – peaches, pears, apples, etc., – all fenced in with a regular quickset hedge, sweet-brier, and blackberries. The owner made £500 a year by that garden. Mrs. Marsh, a sweet old lady, was most hospitable and kind, bringing us a delicious tea into our rooms with her own hands, and leaving us to rest in perfect peace after all our jolting. She sent us on in the buggy to Bendemere – a pretty green meadow with a clear river running through it, bordered by casuarina trees or "she-oaks," so called from the original Indian name of shiock, which had been again varied into "he-oaks," "swamp oaks," and even "oaks" alone, all being species of casuarina, a tree as unlike the English oak as it is possible to find.

The inn was an ideal one, and we longed to stay a week – till the night came, and then we wished ourselves elsewhere. There was "any amount" of fleas and noise: first coaches and supper; then drunkenness and gambling till three in the morning in the bar, which was only separated by planks of wood from our pretty little rooms. I had heard of such unquiet nights, but seldom met them in Australia, and we were not sorry to be off, fortunately getting the inside of the coach to ourselves on that day of constant rain. We descended steadily all the way to Tamworth, where we found a warmth we had not felt before.

From thence the railway brought us on to Musselbrook, over a richer and wetter country, where imported green willows marked the stream's course, and patches of white iris and sweet-brier were growing even more luxuriantly than in their native homes. How it rained when the rain did come!

We did not reach Merriwa till nine at night, when C.C. met us at the little inn and drove us

162

over the deep black mud to his home at Collaroy, on the top of a steep hill, from whence he had a view over the rich cleared country dotted with gum-trees, and winding river, bordered with so-called "oaks," to the Liverpool range. I painted a pretty swallow's nest there, which was built in a tin funnel hanging to a nail in the cellar. When I hung it up again, after taking it away for some two hours, the owners (which were in a terrible fuss to

White Gums (*Eucalyptus* sp.) and casuarinas with a Koala (*Plascolartos cinereus*) in the fork of a Gum Tree and a Platypus (*Ornithorhynchus anatinus*) in the water.

know what had become of it) flew to it at once, and in five minutes time the hen was sitting on the eggs as quietly as before. Some fuss was made about starting again, things being somewhat untidy at Collaroy, roads being so bad, and our two portmanteaux considered heavy (I could lift and carry them both without difficulty); but at last horses were caught, and we started in two buggies with two horses each, and slept at Cassilis, ten miles off. Our two pairs of horses crawled over sixty-three miles of uninteresting scrub the next day, with only one hour of positive rest, and we were glad to get to the smiling valley of Mudgee, with its lovely river and large straggling town, with (strange to say) fine groups of large trees left here and there to shade it.

We were most kindly received by Mr. and Mrs. L. in a really comfortable new house on the top of a hill. The clever old grandfather had been one of the first settlers in New South Wales. The children, as usual, were very nice, and one little boy found me the double nest of the yellow-tailed tomtit – a very dirty specimen, but good enough to paint from. The cock sat on the upper nest to entertain his wife below while she was hatching the eggs, his nest being open to the sky, hers entered only by a small hole in the side. It was not easy to get away; nobody seemed to know how to go, so we went on and on by coach, and on reaching Wallerawang found no buggy, so continued our progress to the railway, where, at one in the morning, we found no rooms. After a weary hour, we started by train up the zigzags to Mount Victoria on the Blue Mountains, which we reached at 3.30. My travelling companion continued on to Sydney.

The rain came down in such torrents that, after a quiet day of painting and rest, I went by rail to Sydney, to stay with some friends of the Torres Straits, Mr. and Mrs. M. I went to see a most agreeable man, Dr. Bennett who was about eighty, but took me all over the Botanical Gardens, showing me with great delight how much bigger the *Ficus bennettii* was than the *Ficus moorii*. The gardens were lovely, but I longed for the country, and escaped to Camden, thirty miles off, by rail and road – one of the oldest settlements of New South Wales, and certainly the most lovely garden in Australia. Three generations of Macarthurs had devoted themselves to it. The present Sir William

had spent thousands on its orchid-houses, and had exchanged plants with every botanical garden in the world. I shall never forget my first walk in that garden. The verandah which ran round the house was one mass of blooming blue wistaria; close by were great jubaea-palms from Chili, a monster I had never seen before. There were quantities of Japanese and Chinese plants, and quite a grove of camellias in full bloom, strawberries with ripe fruit, lemons, bananas, apples, figs, olives, every variety of climating contributing to fill that garden. There were acres of bulbs and different herbaceous plants scattered about the park in different directions by themselves in unexpected places, and large vineyards for wine-making, which I feared would not be kept up when the old gentleman died.

They lent me a buggy with a fat horse and driver for a week, and I went through pretty scenery till I reached the top of the Illewong Mountains, and went down the wonderful bit of road to Balli. At the top I saw many specimens of the great Australian lily or doryanthes, but they were not in flower. I watched a spike of one, seven feet high, off and on for two months at Camden, and it never came out (the one I afterwards painted at Kew took five months after it had begun to colour before it really came to perfection). Another day I stopped to paint a gigantic fig-tree standing alone, its huge buttresses covered with tangled creepers and parasites. The village was called Fig-tree village after it, and all the population was on horseback, going to the races at Wollongong.

At the lake of Illawarra we again found ourselves in the tropics, all tangled with unknown plants and greenery, abundant stag's-horns, banksias, hakea, and odd things. The road up the Kangaroo river and over the Sassafras mountain is

Wild flowers of the Blue Mountains, New South Wales. Conspicuous in this collection is the star-like inflorescence of the Flannelflower (*Actinotus helianthi*), while lying in the left foreground is a dark blue *Patersonia* and a rose-flowered *Boronia*. Other species represented belong to the genera *Pimelea*, *Diuris*, *Eriostemon*, *Epacris*, *Correa*, *Kennedya*, *Daviesia*, *Helichrysum*, *Lambertia*, *Styphelia* and *Tetratheca*.

pretty. After turning the top of the hill we came suddenly on the zamia or cycad – a most striking plant, with great cones standing straight up from the stem. When ripe the segments turn bright scarlet, and the whole cone falls to pieces, then they split open, and show seeds as large as acorns, from which a kind of arrowroot can be extracted, after washing out all the poison from it. The natives roast and eat the nut in the centre of the scarlet segments.

After my return to Camden the railway took me up into the mountains again to stay with the wife of the Prime Minister, a man of great talent, who delighted in collecting beautifully bound books, exquisite china, and other nicknacks to fill his pretty house. But he seldom lived there, leaving his wife and youngest daughter to keep house in almost complete solitude; and perhaps it was that very busy and unselfish life which made the daughter so attractive to me, and such a delightful companion. She knew so much of the plants and birds and beasts around her, and loved the beautiful views over the sea of blue forest and real sea beyond as much as I did.

I found twenty-five different species of wildflowers in ten minutes, close to the house, and painted them. The garden was cut in terraces, descending into the real virgin forest, with fine gums and banksias left standing amongst the imported flowers. One could hardly see where the wild and the tame joined. The native pear was in both fruit and flower. I painted it, and also a pretty little kangaroo-rat which Lily had had as a pet.

On leaving them I returned to Sydney, to the fine old house of Sir G.M. at the head of Elizabeth Bay, which was then occupied by his nephew and his wife. I did not care for the town, and was glad to be in the fine old comfortable Queen-Anne sort of house in such a miserable drizzling rainy weather. Here I could work on quietly; but I returned for a night to Camden to pick up my box and then was picked up by an express train and stowed away on a shelf in a real Pullman car. We reached the end of the rails at eleven next morning, and I was crammed into an omnibus with thirty other persons, and with nine horses to drag us. The party were all most good-humoured, and the country was richly cultivated, with orchards and corn. It was Sunday, and they deposited us at a place called Albury.

Above: Koalas (*Phascolartos cinereus*) and a Wooden Pear (*Xylomelum occidentale*).

Opposite: View of Sydney Harbour with a Wonga-Wonga Vine (*Pandorea pandorana*) and Crimson Bottlebush (*Callistemon citrinus*).

A Vision of Eden

We started on Monday at six o'clock with the same troop of good-tempered people, and crossed the fine running river Murray, the one big river of Australia. Three miles brought us to the end of the Victoria railway, where my free pass ended, and I had to pay like an honest woman again.

As we got nearer Melbourne we saw miles of pasture, covered with a kind of coloured dandelion (*Cryptostemma calendulacea*), whose seed was brought over from the Cape only a few years before and now grew everywhere; but it did no harm, for the cattle ate it. Dear old Mr. C. met me at the station at Melbourne, and drove me home in his nice close carriage with splendid horses to the most comfortable of stone houses, with two storeys and a lift, as well as a beautiful marble staircase. With those kind people I stayed whenever I was at Melbourne, in the greatest peace and comfort and perfect quiet.

Melbourne is a noble city, and its gardens are even more beautiful than those of Sydney, with greater variety of ground, and lovely views over the river. It is by far the most real city in Australia, and the streets are as full of quickly-moving people as those of London. Baron von M., the great German botanist, gave us a great deal of his company at Melbourne. The Great Exhibition was going on, but as tiring as those things generally are, and I did not often go into it.

Mr. C. put me into a coach one morning at eight o'clock. It was crowded, but I had a decent kind of woman next to me, whose husband was a Chinaman! I made friends with both. When we came to Healesville we changed coaches for the second time. At Lilydale I had a talk with a magpie in a cage, and the master came and showed it off – a real wonder of a bird. After leaving Healesville I got on the box-seat, and saw the lovely forest as we mounted the steep ascent. The driver said he did not believe any of the trees were 320 feet, and that they could make the Baron believe anything they liked; but it was a noble forest. The trees ran up like gigantic hop-poles, with thousands of tree-ferns under them, also straight, and thirty feet high, swelling much at the base of their stems, a nice undergrowth of young gums and other shrubs under them again. The little inn at Fernshaw was perfect quarters, with a lovely little garden of sweet flowers, surrounded by the forest, and with two nice girls and their brothers to take care of me, their only guest.

It is difficult to realise the great height of those gum-trees, they are all so much drawn up. It was two and a half miles up hill to get to the tallest group, and was very cold, with some rain. I was glad to warm my half-frozen fingers by a fire in one of the blackened tree-stumps now and then. Not a creature passed me all day, and there was no noise, except the songs of birds and the jeerings of the laughing jackass. The leaves of those amygdalina gums were much larger and darker in colour than the other sorts I had met with; its young shoots were copper-coloured, and the stems were just peeling off their old bark, showing all sorts of delicate gray and red brown tints. The tree-ferns (chiefly dicksonias) were unfolding their golden crowns of huge crooks. Every step brought me to fresh pictures, but it was impossible to give any idea of the prodigious height, in the limited space of my sheets of paper. The Baron had said, "One thing I must entreat of you, Mees; when you will go to the forest, make a boy go before you and beat about with a stick, and please, you will always keep your eyes fixed on the ground, for the serpents are very multitudinous and venomous." But the girls told me the only place a snake is ever seen is on the high road; there it is dangerous. Snakes here only like dry places.

I had a delightful day returning. The coachman started by picking up a lot of school-children, one after the other, with their slates and dinner-baskets, nearly filling the inside with the happy little creatures; then he tossed me the reins and jumped down and into the bush after a snake,

Wild flowers of Albany, West Australia. In the foreground, among others, are: *Anthocercis viscosa,* the large white flower; *Thysanotus* sp., purple flowers with fringed petals; *Leschenaultia biloba,* deep blue flowers; behind is *Burtonia conferta.* Hanging in front of the base are *Kennedya coccinea,* with the elegant blue *Sollya fusiformis* on the right and the white and pale pink inflorescence above, somewhat like a Maltese cross, is *Xanthosia rotundifolia,* intermixed with *Pimelea rosea.* Above these is a species of *Petrophila,* with the pink, hop-like inflorescence of *Johnsonia lupulina.* Behind and to the right are several species of *Stylidium,* with a dark purple-brown *Tetratheca filiformis* in front.

which however escaped him. We passed large vineyards, but coachmen do not think much of a wine which only costs a shilling a bottle. They also had hop-gardens, which answered well. It is curious how we have introduced all our weeds, vices, and prejudices into Australia, and turned the natives (even the fish) out of it.

I only stayed a night at Melbourne on my return, and then went on board the great P. and O. boat, *Malwa*, and on by it to the harbour of Adelaide; but the sea was very high, and though I had written to warn Dr. S. that I would land and see his famous gardens, I gave up all thoughts of doing so when I saw how it was to be done. It was fearfully cold, and we were all glad to reach the calm and sunny waters of King George's Sound, and to land on the sandy shore of Western Australia.

Mrs. R., the flower-painter I had heard so much of, sent her friend, the young manager of the bank, to meet me on board, and to bring me to the little cottage she was lodging in, where she had kept a room for me, and at once introduced me to quantities of the most lovely flowers – flowers such as I had never seen or even dreamed of before. The magistrate, Mr. H., came soon after, and wished me to go on to stay at his house, but I was too well off to move. He told me the only way of going to Perth was either by the horrid little coasting steamer once a fortnight, or by the mail-coach, which also went once a fortnight, travelling day and night, with passengers and boxes all higledy-piggledy, any quantity in a sort of drag or open cart. It generally broke down and killed one or two people. If I hired a private carriage, it would cost me £25 for it alone, without the horses. I said "Thank you," and wrote to the Governor by the mail just starting, who telegraphed in reply that he would send me a carriage at seven shillings a day hire, and I might have the free use of police-horses and a driver as long as I stayed in Western Australia, to take me wherever I wished. Long live Sir H.R.!

Flowers of the West Australian shrub Kangaroo-paw or Kangaroo's Feet (*Anigozanthus flavida*) and Bottlebrush (*Callistemon speciosus*), with on the left side a branch bearing seed-vessels.

So I stayed on at Albany till the carriage came, and found abundance to do. The garden of our little house led right on to the hillside at the back, and the abundance of different species in a small space was quite marvellous. In one place I sat down, and without moving could pick twenty-five different flowers within reach of my hand. The banksias were quite marvellous, their huge bushy flowers a foot in length, and so full of honey that the natives were said to get tipsy sucking them.

At last the carriage came, carrying its own wheels inside, and having substitutes, which did not fit, in their place. The head of the Albany police assured me the axle was broken, and no one in Albany could mend it; the carriage must go back by sea, and I might hire another carriage at fifteen shillings a day, etc. I said, "No, I would rather give up Perth, and go back to Victoria by the next steamer." After an hour or two he came back to say they could put the back wheels on the front, and the front ones back, and he thought it might be ready in three days. I grumbled and growled at so much delay, and was next told it would be ready to start in the morning, if I liked, which I did, and started, passing wonderful things, though the best flowers were said to be over. The chief excitement was a group of hakeas, like a tall hollyhock with leaves like scallop shells, perfect cups growing close together round and round the stem. Every leaf had a flower or seed-pod resting in it; the flowers were pink, but the chief peculiarity was, that every spike of leaves was gradually shaded downwards – the leaves at the top salmon pink, those next yellow and orange, and so into brightest green, blue-green, and purplish gray. We stopped to rest the horses near a large mere, where I found beautiful lobelias, utricularias, rushes, and plenty of boronias and other common Australian plants. After that we wandered over long tracts of sand, wearisome, except that it gave one time to see the endless variety of flowers, for we could only go at a foot's pace. After a while we got on to a better soil, and turned off the road to Mr. H.'s large comfortable house. His wife was quite cross when I said I must go on the next day, though he understood at once that I could not keep the police horses. The first thing I saw when I got up in the morning was my police-driver hammering under the carriage again. We had no

A Vision of Eden

Above: A forest scene from West Australia with a grass tree (*Xanthorrhoea* sp.) in the foreground. On the right is a species of *Kingia* which is probably different from *Kingia australis* in the centre. On the left the cycad is probably *Macrozamia fraseri*, with a species of *Banksia* behind.

Opposite: Two West Australian shrubs – foliage and flowers of *Banksia grandis* and a blue-flowered species of *Comesperma*, probably *C. volubile.*

accident, but the country got drier and drier. All the flowers seemed to turn into everlastings, as if they were determined to fill the gap left by so many other departed flowers, and to keep up a show till the others began again. One sort was especially lovely – a white fluffy ball, with pink satin spikes set in it, and no visible leaves or stalk. It looked like a gem on the white sand. Yellow, pink, and white, all the everlasting flowers were there in quantities, also the strange plants known as "kangaroo's feet." I saw some specimens of the curious gum-tree which grows at the edge of the Marlock Scrub. The latter has turned back many Australian explorers by its density, being almost impossible to penetrate.

We passed only three houses in a sixty mile drive, and could get no food, but my Irish police driver, O'Leary, boiled his "billy" and made some tea at Black River, where the water was worthy of its name. At Rogenut I lodged at a police station, and was so surrounded by policemen calling me "your ladyship" that I felt like the Queen of the

A Vision of Eden

Cannibal Islands, and rather a dangerous character. The sundew grew into perfect little trees near
there, and we passed a mile of everlasting flowers,
one perfect bed of them in the burnt-up grass.
Then we came to another marvellous sandy plain,
and every kind of small flower – great velvety
"kangaroo's feet," with green and yellow satin
linings, exquisite blue or white lobelias, heaths,
and brooms; the latter was very tall, sometimes
bordering the road like a hedge, and whipping one
in the face as the carriage pushed through. We
met, farther on, groups of natives with bundles of
long arrow-headed spears, which they throw at
any animal they wish to kill. I also saw some
sandal-wood (*Fusanus spicatus*) trees, one of the
gums, which has the same scent and qualities as the
real sandal-wood. It grows to the size of an English
apple-tree, and is hung over with a mistletoe
which mimics its own leaves exactly. On that day,
too, I shall never forget one plain we came to,
entirely surrounded by the nuytsia or mistletoe
trees, in a full blaze of bloom. It looked like a bush-
fire without smoke. The trees are, many of them,
as big as average oaks in our hedgerows at home,
and the stems are mere pith, not wood. The whole
is said to be a parasite on the root of another tree
(probably the banksia).

After that we came to William River and from
there had a heavy drag over the sand with tired
weak horses, as the good ones had to be kept back
for the expected mail. We arrived early in the day
at the little inn of Bannister, which stood quite
alone in the forest, with a lovely garden of sweet
flowers in front, and a most friendly landlady. I
had a quiet afternoon at my ease, with her
gossiping at my side, delighted to get a new
talking-post. The next day we had the same tired
horses and a seven hours' drag over the sand. At
two o'clock we got some food and fresh horses,
and came on eighteen miles at full gallop, covered
with dust, holding on for dear life. Many horses in
Australia will either go full gallop or not at all, and
when they are started (no easy matter), the thing is
to keep them at it by hooting, whipping, and
shouting.

I found a note at the next hotel from the High
Sheriff, asking me to come and stop with him; but
I had promised to go to Mr. and Mrs. Forrest, and
met them in the road, they having come fifteen
miles to meet me across the country from

Freemantle. The sea at Freemantle was edged with
delicate little shrubby plants, then out of flower,
but their leaves and twigs had a whitish look, and
seemed to harmonise with the dazzling white sand
in a way that green leaves would not have done. I
painted the flowers of the jarrah or mahogany
gum-tree there; it was loaded with bloom like our
wild cherry at home, and I put in the F.'s pink
cockatoo amongst it.

I went on to Government House at Perth, to
thank the Governor for all his kindness in sending
to fetch me, and then Lady R. made me come and
stay with them. But when I heard that the
Eucalyptus macrocarpa was to be seen in flower at
Newcastle, horses were again ordered for me, and
I was sent over there. We went only too fast
through all the forest wonders, and I screamed
with delight when the small tree came in sight
close to Mrs. H.'s house. Every leaf and stalk was
pure floury white, and the great flowers (as big as
hollyhocks) of the brightest carnation, with gold
ends to their stamens. It was well worth coming
for. The tree had been common enough in old
days, on the edge of the desert, but the sheep had
taken a fancy to it and had gradually eaten it all up,
and they were carefully saving the seeds of this one
that they might sow them and raise up more food
for the sheep!

After a day's rest back at Perth I was packed into
a new carriage – a box on four wheels, with two
seats across and a bit of canvas stretched over four
posts to keep the sun off; no means of mounting
into it except by the wheel, and I was warned to be
always well seated before fresh horses were put in,
as they were apt to run away at first. We passed
through glorious forests of big gums and mahogany trees, and plains of paper-bark trees, with
their curious white twisted trunks and velvety-
green heads, sometimes sprinkled with small
white flowers. We saw also the native pears, with
long bunches of greenish-white flowers, and gray
velvety fruit, the younger ones almost rosy, like
the winged seeds inside. I found quantities of a lilac
satin flower about the size of a primrose, with oily
grassy leaves. Picking it made my fingers wet,

Karri Gums (*Eucalyptus diversicolor*) near the Warren River,
West Australia, with casuarinas and emus in the foreground.

174

though it was growing in the driest white sand, among other dry things, and was said to be *Byblis gigantea*, a sort of sundew mentioned by Darwin, but not yet seen in England. We passed through Pinyarrah and Drake's River and just before Bunbury met a mounted policeman with a kind note from the magistrate and Mrs. C., asking me to stop with them two miles short of the town, in a charming large house by the side of a clear river, with olive-trees, mulberries, and other importations all round the garden.

The country was more English-looking in that remote part of Western Australia than anywhere else that I had been to on the vast island. We went for a drive through corn-fields and meadows with noble red gums isolated like the old oaks at home, with hedges and numerous gates which had to be opened and shut in the same tiresome way as at home. Bunbury is a model place, with a long wooden jetty running out into the exquisite blue bay; at the end of it a ship was being loaded with "mahogany" and other precious woods. After leaving this pretty place we entered miles of sand, and such wretched land that even trees were stunted; only the swamp banksia, with its smooth slate-coloured stem and thin white leaves, was very large, but the orange bottle-brush and some other swamp bushes were in great beauty. Patches of lobelia and other tiny coloured flowers made the sand gay. Some hours ahead Mr. H., the Head of the District Police, had arranged most kindly to go with me on horseback, and to make his tour of inspection fit into my plans. He was a model of an active young English gentleman, and soon after arriving at Vasse he arrived also. Vasse is the chief port of that part of Australia, and the fortnightly mail steamer always touches at it.

From Vasse we had a long drive over deep sand and swamps, a rest, and a change of horses. Just before sunset we reached the great wooden bridge at Blackwater in the midst of a splendid forest; the police station is in a lovely situation with sweet flowers growing round it, and has a very nice master and mistress, who made us most comfortable. The next night was at a house of greater pretension but less comfort, belonging to the governor of Honduras. We started at daylight again with considerable difficulty, one horse declaring he never would, should, or could go up-hill. The other had the same aversion to going

down; he went sideways, sat down, and had to be held up, while I led the three loose ones. O'Leary lost his head entirely, and soon after that I found myself on my knees in the back of the carriage, and saw him, the horses, and the front part of the carriage, going off separately. Mr. H. and his famous police-constable of Blackwater cut down some young trees, and patched the vehicle together. We started again, but O'Leary's nerve was hopelessly gone; he ran first into one tree and then into another, and finally completed the destruction of the carriage within a mile of the waterfall where we were to have rested, and just at the entrance to the forest of perhaps the biggest trees in the world. It was eleven miles from "The Warren," the place of all others I wanted to go to, but which the Governor had said was impossible; and I thought it a most lucky accident when Mr. H. decided on riding on and begging its owner to come to the rescue and take me there. I spent four delightful hours sketching or resting under those gigantic white pillars, which were far more imposing than the trees of Fernshaw; their stems were thicker and heads rounder than the amygdalina gums.

About five o'clock Mr. B. came (a cousin of my cousins of Beechboro'). He drove a heavy kind of drag, and I felt I had no more fear, his driving up and down and in and out of stumps and trees was so sure, going at a good trot all the while. We were under enormous trees, chiefly white gums, as smooth as satin, and sometimes marbled, with a few rough red trunks or "black butts" among them, and small casuarinas and shrubby bushes underneath. The Warren, a rambling, untidy house with farm-buildings, stood near a clear river, in a hollow, with two fields surrounded by forest and "ringed trees"; nothing ever seemed to have been repaired there. How could it be, with no servants and no neighbour within thirty miles? My old carriage came on after we arrived, patched together so that it might be sent back to its owners at Perth, but not fit to go on with; indeed, they all said O'Leary's nerves were far too shaky for any one to trust him as driver in future. So Mr. H. went on his tour of inspection eastward, and I turned west, with Mr. B. in the heavy drag. We found a lighter carriage at Mr. B.'s, and went on for the night to the pretty cottage of Blackwater again, with no adventure but Mr. B.'s nose taking

Second Visit to Borneo – Australia

View from the Botanic Gardens, Hobart Town, Tasmania, with in the foreground grass-trees (*Xanthorrhoea* sp.) and an Oyster Bay Pine (*Callitris tasmanica*).

to violent bleeding. We reached Vasse the day before the steamer started, and it took me back to Albany round the stormy West Point of Australia.

At Melbourne I was again most kindly welcomed by the C.s in their comfortable home, and even Baron von M. was excited over my paintings of the nuytsia and the *Eucalyptus macrocarpa*, which he had named, but had never seen in flower. When I showed him the bud with its white extinguisher cap tied over it, which I was saving for Kew, he said "Fair lady, you permit I take that?" and calmly pocketed it!

1881. – On the 20th of January I crossed the Straits in a small steamer, outrageously crammed, and every one sick except myself: it was most horrible! We stuck in the river near Launceston, and a friend of a friend's friend came to meet me, but was so busy that I told him to go on shore again, waiting myself till the steamer reached the quay, when I got on shore, and on by rail to Deloraine, where he had telegraphed to the parson to meet and lodge me. Mr. E. and his wife were charming people, and really liberal. The bishops at home had refused to ordain him, so he had it done in Tasmania or Victoria. The country was not in the least attractive to me; it was far too English, with hedges of sweet-brier, hawthorn, and blackberry, nettles, docks, thistles, dandelions: all the native flowers (if there were any) were burnt up. One lovely flower I heard of and was taken a long drive to see. It was – a mullein!

After three days' rest I came over eleven hours of slow railway across the island to Hobart. Here Mr. S., the clever school-inspector, had taken a room for me in a boarding-house at the back of the town, near his own house. I was not allowed to stay there, though I had no formal introduction to the acting Governor, Sir H. Lefroy. When I went

A Vision of Eden

A selection of flowers from Mount Wellington, Tasmania. In the foreground on the left are the clustered red and bluish berries of *Cyathodes glauca*, the rosy flowers of a *Pimelea*, the lilac flowers of a *Prostanthèra*, and behind these the small spicate flowers of *Bellendena montana*, the red berries of *Lissanthe strigosa* and a Tasmanian Ivy (*Pseudopanax gunnii*). Above is a branch of *Eucalyptus cordata*, bearing seed-vessels. To the right is a cluster of Tasmanian Lilac or Victorian Christmas-bush (*Prostanthera lasianthos*), followed by the large white flowers and ruddy foliage of *Anopterus glandulosus* and the crimson flowers of *Telopea truncata*. The lobed leaves behind the *Anopterus* are those of the Celery-topped Pine (*Phyllocladus asplenifolius*). Hanging in front are the showy blue-fruited *Billardiera longiflora* and a white-fruited variety of *Drymophila cyanocarpa?*, with two or three native Cherries (*Exocarpus cupressiformis*) at the bottom.

to inquire for letters he and his wife received me like an old friend, and insisted upon my coming at once and making Government House my home while in Tasmania. They took me round their garden to see the lovely views over sea and river, and were full of plans for my seeing the Island, which, however, were never carried out. They had a tree loaded with dark apricots like purple velvet, crimson inside, but tasting as other apricots do: it was very beautiful. Cherries, raspberries, every kind of fruit which grows at home grew better than at home. Half the jam in the world is made in Tasmania. It is sent on to the colder parts of New Zealand and Australia, where enormous quantities are always consumed.

Before I went to stay at Government House, the Head of the Gardens came and drove me up the Huon road to the shoulder of Mount Wellington. Four miles of walking took us to the lovely spot where the clear water bubbles out amongst the fern-trees and all kinds of greenery. After a rest we plunged right into the thick of it, climbing under and over the stems and trunks of fallen trees, slippery with moss, in search for good specimens of the celery-topped pine, of which we found some sixty feet high. It was not in the least like a pine, excepting in its drooping lower branches and its straight stem: the leaves were all manner of strange shapes. We also saw fine specimens of Sassafras (which yields an oil rivalling the real American Sassafras in value), and the dark myrtle or beech of Tasmania, and quantities of the pretty, pandanus-looking plant they call grass-trees or richea, really a sort of heath. The whole bunch looks like a cob of Indian-corn, each corn like a grain of white boiled rice, which, again, when shed or pulled off, sets free the real flowers – a bunch of tiny yellow stamens, with the outer bracts scarlet. The famous blue gum (*Eucalyptus globulus*) was rare even there: strange that this should be almost the only species known or grown in Europe.

Rainy weather came on, and I was glad to be in the beautiful rooms at Government House, my room looking over the blue bay with the great terrace of flowers in front. Miss L. and myself were taken one day by the A.D.C. to an old barber's in Hobart, who collected and stuffed skins, etc., and he showed us all his treasures; the greatest being three little opossum-mice, the smallest of all

marsupials, with prehensile tails – soft little balls of fluff with very big ears. They were only lively at night, but quite tame and easily managed in the day-time. The old man almost cried at parting with them, but at last I persuaded him to do so (for a con-si-de-ra-tion).

I went up the Derwent in the steamer to New Norfolk. The river spread itself out like a series of lakes, with rocks closing it in where it narrowed, all arranged in horizontal strata like walls of gigantic masonry. It ran through a rich bit of country full of hops, and orchards loaded with fruit. Hedges of hawthorn loaded with red berries, sweet-brier, and blackberries, – all was too English, – it might have been a bit of Somersetshire, as I drove along the beautiful river-side for four miles to visit Mr. R., an English squire-farmer. His little house was smothered in flowers and fruit, the windows so darkened by them that inside one could hardly see to read. On Monday I was sent down to sleep at the inn, so as to catch the first steamer in the morning to Hobart, where I had ordered a carriage to meet me and take me over the Huon road, which creeps round the shoulder of Mount Wellington, commanding lovely views of sea and island. The forest scenery is of the grandest description, with undergrowth of tree-ferns and shrubs of many kinds, that called the Tasmanian lilac (*Prostanthera lasianthos*) being the most striking.

I stopped at a nice little inn by the long bridge on the Huon river, and drove on again the next morning along its lake-like banks, past Franklin to Geevestown, where a whole family had settled and populated a district, establishing saw-mills and a tramway far into the noble forest they were gradually destroying. Then I drove back to Franklin, where the good rector and his wife kept me, sending back for my portmanteau from the bridge inn. The finest tree-ferns I had seen in Tasmania grew there; many of them were hung with clematis, like the English one. Franklin was a damp feverish swamp. The winters were very long, and strangers seldom came to cheer them. It had been an old convict-settlement, and the place and people had a bad name; wrongly, for they were a most sober, peaceable community.

I had ordered a room, as I passed up the shoulder of Mount Wellington, at the Red House, and returned to it on my way back. It had been built by a man of taste, years ago, for the sake of the view. He died before it was finished, and it was now occupied by a widow with a large family of nice children, who made it delightful quarters to stay in.

View in the forest on Mount Wellington, Tasmania. In the foreground are tree-ferns (*Dicksonia antarctica*) and *Richea dracophylla*. In the centre is a large specimen of a southern beech *Nothofagus cunninghamii*, with on each side Celery-topped Pines (*Phyllocladus aspleniifolius*) and Sassafras (*Doryphora sassafras*).

CHAPTER·X

New Zealand and the United States
1881

SIXTY HOURS AFTER TIME my ship arrived, and I had to leave a large dinner-party at Government House to go on board, where, thanks to my kind friends, I found a cabin kept for me. It was frightfully cold, and I huddled myself up in my opossum rug and read Miss E.'s new novel and Disraeli's. My poor little mice were half frozen, and cold as apples; but I took them in my hands and rolled them about till they got life into them, when they began to yawn and stretch themselves again. Major D. and his wife persuaded me to let them take the poor little things on to Wellington by sea, where it would be warmer, to save myself the trouble of carrying them about overland. No wonder it was cold, with no land between us and the southern ice-cliffs! It was also foggy, and we could not get up to the bluff till past eight in the morning. But the mountains began to show before we landed, and the Southern Island of New Zealand looked fine in the distance: but for the cold I should have liked a month there. A little snow was nestling in the hollows of the rocks, and I felt it was useless for me to attempt it. The near shores were rough and rocky, covered with ti-trees, with many odd things amongst them of the dracaena kind, the oddest thing being the *Panax crassifolium*. In its young state it resembled the skeleton of a half-folded umbrella; some three or four years later, it would become a round-topped tree with five-fingered leaves.

The train had waited for us, and now took us slowly and with many stoppages over the rich

Dracophyllum traversii from New Zealand.

marshy level ground to Invercargil. Masses of the native flax, with great black bunches of seed pods on tall stems, and now and then a rag of red flower left on them, were all over the plain, as well as coarse grass like the tussock. Invercargil is a widely spreading place, with houses standing fifty yards apart, and has one of the cleanest and most comfortable hotels I was ever in. But the wind howled and the dust blew, and all was cold and dreary to me. I was glad to leave at seven the next morning for real country and new weeds. My train only went as far as Elbow, where I was turned out again for six hours. The air was thick with thistledown, and the native weeds were being stifled by Scotland's royal flower. We passed over miles of grand scrub, every tree new to me (and no gums!), mostly of the "ti-tree" sort, perfect cushions of green velvet of different tints on queerly twisted trunks, whose branches sprang out at unnatural angles, giving a top-heavy look to the trees. Elbow has the great mountains all round its yellow plain, with a clear river running through, and quantities of the tall flax all over it. I settled myself to make a sketch in the glaring sunshine, and found it really hot. At four o'clock we again started through the wildest mountain country, all stones, with rabbits running over them up the steepest places in such numbers that the very stones seemed to be running too. They are becoming a real plague on the island. We descended by two zigzags to the edge of Lake Wakatipu, just as the sunset threw a great purple shadow up to the top of its rocky walls.

A steamboat soon took us farther into the

A Vision of Eden

Above: View of Mount Earnshaw from the Island in Lake Wakatipu, New Zealand. The white-flowered trees belong to the genus *Cordyline* and on the left is a specimen of *Pseudopanax crassifolium*. The tufted plants in the foreground are spear grasses (*Aciphylla* sp.).

Opposite: View of the Otira Gorge, New Zealand. The conspicuous plant with white plumes in the foreground is a Toe Toe Reed (*Chionochloa conspicua*) with, on the rocks, trees of *Dracophyllum traversii* and one of *Metrosideros umbellata*. The white-flowered plant nearest the front is *Ranunculus lyallii*. On the rocks behind are tufts of the Kid-glove or Leather Plant (*Celmisia coriacea*). The tree at the top on the right with light-green foliage is *Dacrydium colensoi*.

darkness, between high bare mountains reminding me of the Lecco end of Lake Como. Queenstown is on one of the few bits of flat in the whole of the wild lake, a kind of delta brought down by a small stream in flood-time. It is full of summer villas and hotels.

I got very chilled one day when trying to paint, so went for a walk up and down the edges of the cliffs for four miles and back. It was very beautiful, every crack having running water, and the scrub was full of interest to me. The only biggish trees were a kind of beech like that in Tasmania and Victoria. There were many berries, but no flowers; and I felt the happier for it, as the scenery itself was enough to study at once. The wind was bitter, and the waves beat on the shore like the real sea. I felt I could never rough it in such a climate, and my aching limbs could not crawl fast enough to warm me. I sat and wondered if I should ever get home to England and see my gallery finished. Mr. Fergusson had written me that the shell was nearly ready, and I longed to be there without the trouble of going.

A Vision of Eden

On the 1st of March I left Queenstown for Dunedin which is a well-situated town, with hills behind and bays in front; but the hotel I had been told to go to, in the suburb of Leith, was miserably depressing. Everything seemed poor. I tried to walk to real natural country, but only tired myself and my poor aching bones, and I was glad to get on again and further from cold. The views I had heard so much of I never saw, though I incessantly looked, and I was half dead and starved when we reached Christchurch and its well-regulated and extra-English hotel. My box, which I had left at the station of Invercargil to be forwarded on the day I landed, had never turned up or left Invercargil, and I was sick of everything belonging to that cold, heartless, stony island.

The sight of Judge I. and his wife warmed me; they were so thoroughly genial and alive. They knew everybody and everything, and drove me to the club to inquire for my cousin, John Enys, who came up just at that moment, swinging a pot of apricot jam in his hand, which he was going to take back as a treat for me, to celebrate my visit to his station. He took me four hours of rail westward the next morning. He had hired a horse and buggy, which met us at the end of the railway, and we drove over the dreary burnt-up hills to his house, which was called Castle Rock, after a pile of strange old rocks near; they looked like the remains of some fortified place. His own quarters were at the edge of a black beech forest, which gave a more cosy look to the spot, but they were a hard, cruel sort of trees, the very tallest not more than forty feet high, with leaves as small as the box, under which no green thing liked to live; their branches feathering to the ground like Cedars of Lebanon. John corresponded with all the scientific people of those parts, and got into the wildest excitement over a new weed or moth. He sent a man up the hills some 2000 feet, and had some large specimens of the vegetable sheep brought down (*Raoulia*) – a mass of the tiniest daisy plants with their roots all tangled together, and generally wedged in between two rocks, leaving the surface of the colony like a gray velvet cushion: at a distance the shepherds themselves could not tell it from the sheep they had lost.

We rested a night, and then drove on so as not to waste the fine days, crossing miles of weary waste and some lovely tarns with reflections clear as in a looking-glass. We had luncheon in a nice little inn, and then crossed and recrossed the wide river-bed of almost dry stones, a hard pull for the horse. After that came a long mount through the dry beech-forest; till at the top we reached a completely different scene, and a vegetation which rewarded us for all our trouble. The most remarkable plant was a tree which looked something like a small-leaved dracaena, or screw-pine, but which was really a heath! All its under leaves were purple, and its stems bright salmon-colour; the flower was over, but the terminal spikes remained, and were also purple. The whole was one of the most curious growing things I had ever seen in any land. The whole gorge was lined with small shrubs of every tint and shape. It was very narrow, and would have been perilous with any but quiet horses; but ours was most tractable. I found a bit of edelweiss very like the Swiss one, and a few remaining flowers on the kata – a parasite tree which had then quite coloured the hillsides with deep crimson. It seemed to delight in hanging at right angles from the rocks, and in the most inaccessible places. At the bottom of the pass, where two streams unite, we found a comfortable little wooden inn. I did my best to sketch in a wood carpeted with todeas and other ferns, in spite of a tremendous gumboil and a swelled face, and a wind which whistled through the valley, making the whole house rattle.

John said we should soon get into a country of bare stones again if we went beyond this famous Otira Gorge; so, after two nights there we turned up the hill again, over the stony river-beds, through driving rain and hail and furious wind, and took refuge at Mr. B.'s farm or "station," as it was called. Mr. B. kept cows, – a great luxury for

New Zealand flowers and fruit. The spherical plant in the foreground is a small specimen of the Vegetable Sheep (*Raoulia eximia*) which is said to resemble a recumbent sheep. Behind are fronds of the fern *Trichomanes reniforme* and hanging from the vase, the prickly leaves of the Southern Bramble (*Rubus australis*). The yellow berries belong to the Karak (*Corynocarpus laevigata*) and are edible. On the left are some hanging blue spikes of *Hebe speciosa*, whilst the centre is filled with crimson flowers of *Metrosideros tomentosa* contrasting with the white-flowered *Hohera lyallii*.

those parts, – and sent my cousin a potful of butter by every mail, while he supplied them in exchange with his surplus newspapers. When the rain had a little abated we drove back to Castle Rock Station, and the next morning found snow all over the near hills, and the ground white with frost all round. It was far too cold to sketch, but I enjoyed a day among my cousin's curiosities. His little room was crammed with them. He showed me several eggs of the kiwi; one sort was sky-blue, and nearly as big as the eggs of an emu. No wonder the bird was said to die after laying it. If it died, who sat on it? Had it a stepmother? We drove through torrents of rain to the station, and the mud splashed the very top of my hat; but we had four hours to dry in before the train started.

At Christchurch I stayed with Judge I. and his wife, and found the box, which had taken three weeks coming from Invercargil at the other end of the island. But rheumatism was getting more and more possession of me; and we crossed in the big ship *Rotomahana* to Wellington, the capital of New Zealand, where the wife of the Premier received me most kindly. I was ill and miserable, though I tried to work still, going by railway three miles along the shore, and then crawling to a garden (which had once been made by a man of taste, and now was used by a nurseryman). Here I got good studies of the nickau palm, the most southern of all palms.

I was quite glad when the Governor came, and I moved into his house and heard him abuse the island and all belonging to it with as much heartiness as I did. He said, justly, something must be wrong with a country which required so much laudation. I left Wellington on the 27th of March, and reached Auckland on the 29th, stopping on the way at Taranaki, under the shade of Mount Egmont, a perfectly formed volcano of a steeper slope than any I had before seen.

At Auckland I boarded the *Zealandia* and was put at table with a set of third-class colonists, who scrambled for the very indifferent food like pigs. The Equator nearly cured my rheumatism, as I

had expected, though I could not walk without difficulty for two months more.

San Francisco, where I landed on the 20th of April, 1881, was in a terrible whirl and noise, and the Palace Hotel, at thirty shillings a day, was quite perplexing in its vastness. I brought back my rheumatism by wandering about the windy streets. I was introduced by a Dr. B. to a Bohemian wood-merchant, who told me the finest redwood sequoias were on his place near Guerneville, many of the trees 200 to 300 feet high. First I took a through ticket to New York for £30 with leave to take twenty years over my journey, if it pleased me. Then before going to the redwoods I moved to Judd's capital hotel at Oaklands which cost only 10 shillings a day. I was still rheumatic, and very lazy about starting for the redwoods. But I left my opossum-mice to the care of the housekeeper, and departed. The bay was most lovely as the sun rose, driving off the smoke of the big city, and I felt a new creature when I got there. A few steps took me to the other ferry-boat, where I got a good cup of coffee and bread and butter, and was pressed to take eggs, fish, etc. Getting into the train at St. Quintin, and passing the pretty suburb of San Rafael, with its gardens of figs, vines, and gorgeous flowers, we went through meadows blazing with patches of colour like the beds of annuals at home (only fifty times as large): nemophilas, lupins, eschscholtzias, deep blue lark-spurs, pink mallows, sunflowers, etc. I changed cars twice, but had not to wait for them.

We reached the redwood forests all of a sudden, and the railway followed the Russ river through them up to Guerneville, a pretty wooden village with a big saw-mill, all among the trees, or rather the stumps of them, from which it has acquired the common name of Stumptown. The noble trees were fast disappearing. Some of the finest had been left standing, but they could not live solitary, and a little wind soon blew them down. They had a peculiar way of shooting up from the roots round the stumps, which soon became hidden by a dense mass of greenery, forming natural arbours;

Opposite: Foliage and flowers of the Californian or Mountain Dogwood (*Cornus nuttallii*) and Rufous Humming Birds (*Selasphorus rufus*).

Overleaf: Flowers of North American trees and shrubs – the Tulip Tree (*Liriodendron tulipifera*), False Acacia (*Robinia pseudoacacia*), Mountain Laurel (*Kalmia latifolia*) together with varieties of *Rhododendron*.

and many of the large old trees were found growing in circles which had begun that way: a habit peculiar to that tree.

The little inn was capital; and all the gentlemen of the place dined in their shirt-sleeves, and were much interested in my work. They told me how to find the biggest trees, but every one was busy, and not a boy was willing to act as guide or to carry my easel. There was no difficulty in finding the trees; only in choosing which to paint, and how to get far enough away from such big objects as to see the whole of any one. Fifteen feet through, at a yard from the ground, and two hundred or nearly three hundred feet high, were the measurements of the largest. Nearer the river it was prettier and more airy, and there I settled to sketch, in the shade of the young shootings from an old stump. There was an undergrowth of laurel and oak, and many pretty flowers: pink sorrel, trillium, aquilegia, blue iris, and a deep pink rose. I got back to Oaklands after the supper hour (eight o'clock); but the porter brought me a large plate of crackers and butter, a tumblerful of the most adorable iced mixture, and a straw to suck it through.

A day or two after that, I started again with my luggage. I was two hours too early for the ferry. When at last we had started and got to land again, we went for a whole day through a rich corn and grass land, getting warmer and warmer, every now and then passing gorgeous masses of wild-flowers, till we reached the hot plain of Madera, where I had been told I should be dazzled by the flowers, and found them all dried up and turned into seedpods and straw! Madera was the end of the new road to the Yosemite, and I met two fellow-passengers from the *Zealandia* returning from it, who persuaded me to go on there the next day. Coach-travelling was more than usually provoking at that season of flowers, as the coach never stopped in the spots where they grew. I saw at one place a great upright white datura growing amongst the rocks; at another place masses of "prickly elm," with wreaths of yellow flowers. All the flowers grew together in colonies of one species.

It was dark when our aching bones were taken down at Clark's. The old house I had been in before had been burnt down. A two-storeyed house had taken its place, and the bills had grown longer; but it was most comfortable, and nice resting quarters. A road had been made for a coach and four to drive to the Mariposa Grove of big trees. One of them had had a great bit of its inside taken out, so that the coach and four could drive through it! I did not submit to such barbarism, but sat still and painted the snow-flower – a gorgeous parasite of the purest crimson and white tints, which grows at the roots of the sequoia, about 5000 or 6000 feet above the sea. I stopped at the first inn in the Yosemite valley, a homely quiet house, and wandered round it on foot for three days, making no expeditions, but enjoying the grandeur of everything in perfect quiet; and a nice old gentleman of Philadelphia, who had come in the coach with me, brought me wonderful flowers from the mountains above.

The first view of the valley struck me more crushingly even than the first time I saw it; and when I talked of walking back to paint the view I found it was seven miles off! It looked so near! The falls were full of water; they had been dry when I saw them before. My old friend from

Group of Californian flowers. Beginning in front, on the left is a yellow Columbine (*Aquilegia chrysantha*) and behind it the dark blue Common Spiderwort (*Tradescantia virginiana*). In the pot is a large showy white Mariposa Lily (*Calochortus venustus*), with a pink-flowered Centaury (*Centaurium venustum*), a blue-flowered "Chia" (*Salvia columbariae*) and a buff Monkey Flower (*Mimulus glutinosus*). Then come the purple and white, whorled flowers of *Collinsia bicolor*, a large pale blue *Phacelia minor*, the smaller darker blue flowers of *P. grandiflora* and the narrow spike of white flowers is *Antirrhinum coulterianum*. These are succeeded by an azure *Pentstemon azureus,* the more slender crimson *P. centranthifolius* and the plume-like scarlet *Castilleja affinis*. Nestling down among the other flowers are the purple and white clusters of Escobita (*Orthocarpus purpurascens*), with a large yellow *Gerardia* (?), and the yellow star-like flower-heads of Tidy-tips (*Layia platyglossa*) in front. In front of the pot are a dull blue *Triteleia laxa*, a dark blue and white *Delphinium variegatum*, finishing with a showy White Wake Robin (*Trillium grandiflorum*).

A Vision of Eden

Vegetation of the Arizona desert including the candelabra-like Saguaro Cactus (*Carnegiea gigantea*) and whip-like branched Ocotillo (*Fouquieria splendens*) in the foreground. The creeper is probably *Cucurbita digitata*.

Philadelphia rejoiced in the name of Smith. He was very good company when sitting on the verandah of an evening. I was quite sorry to say good-bye and go back to Clark's, where I had another day of rest, painting the calochortus and cyclobothra, then back to Madera.

When we reached the plain a hurricane of wind met us. It was almost as much as the men could stand. I covered myself entirely with my opossum rug, and felt the wind even through that. We had ten miles of it, and arrived just as the train came up; but it wanted its supper as much as we did, and we had abundance of time and food. At 7.30 we halted at Los Angelos for breakfast. The place did not attract me; and I was glad to get half an hour more rail to San Gabrielle, nearer the hills, where I had an hour to wait before the carriage came and brought me through three miles of gardens, both wild and tame, to Villa Sierra Madre. The hotel was absolutely perfect quarters, built as an annexe to the pretty villa and farm of the owner of the land. I reached this delightful place on the 9th of May (quite out of season), and found only one other guest there – an old quakeress, with a piteous cough and nearly blind; but she took a great fancy to me, and sat for hours watching me paint, saying she had enjoyed nothing so much for years. The last night I was there my father-mouse got out, and kept me awake all night. I had to shut all windows and doors, and to block up the fireplaces for fear of losing him. He ran to the very cornice of the ceiling in the moonlight. I watched him in all the highest corners of the room, rushing about like a wild thing, and I knew it was perfectly hopeless to think of catching him till the sleepy hours came at daylight. Then at last I found and captured him, with his head stuck into a hole, and all his body out, no doubt thinking that if he could not see me I could not see him. So I restored him to his sorrowing family, and departed from San

New Zealand and the United States

Gabrielle, whose old convent was long an outside home of the Jesuit missionaries. It now stood almost alone, with the railway-station near it. How the old fathers would have wondered at the noisy steaming thing which carried me on! – over dried-up plains, getting hotter and drier, and more and more flowerless. Often these plains were from two to three hundred feet below the level of the sea, and the still heat was very trying, the ground covered with dazzling sand, with a few scrubby trees loaded with mistletoe and cacti, some like pillar-posts thirty feet high, others branching like candelabra. One shrub (*Fouquieria*), generally called the fishing-rod cactus, – like bundles of spreading fishing-rods united at the base, and bearing a bunch of flame-coloured flowers at their tips, – was most striking.

I had thought of stopping at Zuma, but the confusion and crowd in the unfinished station made me unwilling to leave comfortable quarters in the dark, so I went on through another night "on board," and the next morning arrived at a "Palace Hotel" most unlike the picture in the advertisements! Lots of loafers stood round the door, but the luggage was poked in at the bar-window. I tried all day in vain to get a vehicle to take me to the hills, which I saw miles away. I felt disinclined to stop at other old Spanish towns on that new line. The next morning the train was four hours late, and no one was surprised. It took me across more hot sand-plains of cacti and euphorbias and true Spanish daggers, or yuccas, with great white fountains of flowers on their heads. At Denning the new line ended, and the Santa Fé began. At eight o'clock the next morning we reached the junction with the short line of rail which leads to Santa Fé, the capital of New Mexico, and I have repented ever since of not having had the energy to explore it, but hope if I live to do so. The mountains were covered with pines and cedars, and the ground with patches of lilac verbenas, which became blue as they faded, looking in the distance like water reflecting the sky.

After crossing the high spur of the Rocky Mountains we came down on vast and uninteresting plains, only just recovering from recent floods, with the carcases of drowned oxen scattered about.

I found the hotel in Kansas City so full that, after re-checking my trunk and having some

View in a Redwood (*Sequoia sempervirens*) forest, California, U.S.A.

breakfast, I went on again down the side of the great Missouri through a rich country. The trees were in the freshest spring dress of green: the most elegant wych-elms and willows, solid tulip-trees, exquisite cut-leaved maples, and oaks with leaves five inches long, – the boughs were quite weighed down with them, – hornbeam, ailanthus, hickory, and dog-wood – the latter in flower, but much smaller than the flowers in California. The robinia or locust trees were quite white with lovely blossoms, but they were said to be natives of Virginia. The sides of the rail were bordered by blue iris and tradescantia flowers.

St. Louis is a monstrous city, and its hotel is in character with it. The next morning I drove out to Mr. Shaw's garden, and fell in almost at once with the old gentleman himself, who greeted me like an old friend, and showed me all his chief botanical treasures.

The Mississippi did look big as I crossed its long

195

bridge. I rather longed to spend five days and nights in going down to New Orleans to see the magnolia trees in bloom. I was told they quite poisoned the air with their scent, and made the ground dangerous to walk on, from the oiliness of their fallen petals and seeds. All the way to Cincinnati was park-like and wooded; the city looked grand on its hill in the sunset, over the great river Ohio. But big cities give small pleasure to such vegetable-lovers as myself, so I was "transferred" in an omnibus with four horses to another railway, where we had an hour to wash and feed comfortably, and then found a sleeping-car in a tunnel under the city.

At Washington we were transferred into a train ahead, which rushed on and deposited us in a pretty hotel in a garden, where we had an hour to spend, and a bad dinner, then were again picked up by the original Pullman car and the conductor, who informed me "my mice had just waked up and inquired where I was." Those little beasts were a great delight to my fellow-travellers, and helped to make me a popular character.

The Alldine Hotel at Philadelphia is perhaps one of the best in the world, and the most expensive also. All in keeping with the noble city, perhaps the finest in America. The parks and Zoological Gardens, the Palm-House, and the views from the high ground near them are delightful. I wandered about and enjoyed perfect idleness and solitude for two days, then went on to my friend Mrs. Botta's hospitable house in New York. The first morning she sent for a friend to come to breakfast, and take me out for a hunt after things I wanted. First we went up the "elevated" rail, one which leads from the sea into the country, over one of the principal streets, with iron lace-work resting on iron arches and pillars. We left the rail at the other side of the great Central Park, and walked across some waste ground to the Natural History Museum. We fell in with the Director, and "had a good time." I wanted to make out more about the strange plants of Arizona, particularly the *Fouquiera splendens*, so we went on to Columbia College, and visited first its president, Dr. Bernard, in his den, talking with him through an ear-trumpet and tube: a girl was sitting in a corner doing telegraph work for him. Then we went to Dr. Newberry, who hunted over all his books, finally taking us to the herbarium, where we found not only engravings, but a dried

Rainbow over the Bridal Veil Fall, Yosemite, California, with cypresses, Douglas firs, alders and dogwoods in the foreground.

flower. He took the greatest pains to help me, and his secretary, Mr. B., arranged to go on his half-holiday, Saturday, to hunt for flowers in the woods on Staten Island with me. The ferry and rail took me to his appointed station, where he met me in a buggy, with an old horse to drag it, and we spent a long day driving from wood to wood, and wandering about after cypripediums, magnolias, azaleas, kalmias, andromedas, and other nice things. The sarracenias and dionaeas were also to be found in another part of the island, which was full of pretty villas and gardens belonging to the rich people of New York.

I went on board the *Germanic* on the 4th of June, which reached Liverpool on the 13th of June. My first thought after unpacking was of the building at Kew, and I did not long delay in going there. I found the building finished (as far as bare walls went) most satisfactorily, its lighting perfect. Mr. Fergusson kindly arranged about the decorating and painting of the walls. After that I spent a year in fitting and framing, patching and sorting my pictures, and finally got it finished and open to the public on the 7th of June 1882. I had much trouble but also much pleasure in the work. What need now is there to remember the former? Mr. Fergusson throughout was my best help and counsellor, and towards him I shall always feel the strongest gratitude. I got woods from all parts of the world to make a dado of. Only half of them came with names on them, and half were lost. It was a great difficulty to arrange them, but time mended all. The catalogue I wrote on cards, and stuck them under the paintings; and after I had put down all I knew, Mr. Hemsley corrected and added more information, which he did so thoroughly and carefully that I asked him to finish the whole, and to put his name to the publication.

One bright summer day I went down to Bromley Common to see my father's dear old friend, George Norman, then in his ninety-fourth year. I think he and his dear wife (my best aunt, though no real relation), surrounded by their children and grandchildren in that old house and garden, formed the happiest picture I can think of. They drove me on to Down, a pretty village, and a most unpretentious old house with grass plot in front, and a gate upon the road. On the other side the rooms opened on a verandah covered with creepers, under which Mr. Darwin used to walk up

and down, wrapped in the great boatman's cloak John Collier has put in his portrait. He seldom went further for exercise, and hardly ever went away from home: all his heart was there and in his work. No man ever had a more perfect home, wife, and children; they loved his work as he did, and shared it with him. He and Mr. Norman had been friends for many years, and it was pretty to see the greater man pet his old neighbour and humour him; for with all his great spirit he was very much of a spoilt child, and proud of his age. Darwin seemed no older than his children, so full of fun and freshness. He sat on the grass under a shady tree, and talked deliciously on every subject to us all for hours together, or turned over and over again the collection of Australian paintings I brought down for him to see, showing in a few words how much more he knew about the subjects than any one else, myself included, though I had seen them and he had not. When I left he insisted on packing my sketches and putting them even into the carriage with his own hands. He was seventy-four: old enough to be courteous too. Less than eight months after that he died, working till the last among his family, living always the same peaceful life in that quiet house, away from all the petty jealousies and disputes of lesser scientific men.

(Here follows a short note from Mr. Darwin, written just after this visit, showing his appreciation of my work. The plant referred to is *Raoulia eximia*, a native of the Middle Island of New Zealand, and allied to the Gnaphaliums.

2d August 1881.
Down, Beckingham, Kent.

My Dear Miss North, – I am much obliged for the "Australian Sheep," which is very curious. If I had seen it from a yard's distance lying on a table, I would have wagered that it was a coral of the genus Porites.

I am so glad that I have seen your Australian pictures, and it was extremely kind of you to bring them here. To the present time I am often able to call up with considerable vividness scenes in various countries which I have seen, and it is no small pleasure; but my mind in this respect must be a mere barren waste compared with your mind. – I remain, dear Miss North, yours, truly obliged,

CHARLES DARWIN

CHAPTER XI

South Africa

1882–83

ALL THE CONTINENTS OF THE WORLD had some sort of representation in my gallery except Africa, and I resolved to begin painting there without loss of time. In August 1882 I left Dartmouth in the *Grantully Castle*, a ship historically famed for having once been lent by its owner, Sir Donald Currie, to Mr. Gladstone, for a trip which restored the Prime Minister to health. It took me to the Cape in rather more than eighteen days – one of the shortest voyages ever made. As the ship touched the shore at Cape Town, two friends met me, and put me and my boxes into a hansom-cab. One of them took me all the way out to Wynberg, seven and a half miles, round the western side of the Table Mountain, whose grand crags came down within a few hundred yards of the road, with groves of European fir-trees, oaks, and fruit-orchards, the ground under them covered with white arums, wherever the soil was moist. Australian gums, wattles, and casuarinas were in full bloom, and perfectly at home there. On the sandy flats between the road and the sea were myriads of small flowers: oxalis of most dazzling pink; yellow, white, and lilac heaths, bulbs of endless variety, gazanias, and different mesembry-anthemums. Mrs. Brounger, a most beautiful old lady with silver hair, gave me two rooms in her nice large old Dutch house. Her daughter, Mrs. Gamble, lived close by with her pretty children. Our meals were taken in either house alternately,

Beauties of the swamps at Tulbagh, South Africa – *Watsonia pyramidata*, Red-hot Poker (*Kniphofia uvaria*) and *Richardia albomaculata* subspecies *albomaculata*.

both husbands being away at Government work.

Mrs. Gamble and I had many delightful drives with an old pony, which had a most remarkable talent for standing still. We used to drive him off the road into the thick bush and leave him there for hours, while we rambled about after flowers. The extraordinary novelty and variety of the different species struck me almost as much as it did at Albany in Western Australia, and there was a certain family likeness between them. But the proteas were the great wonder, and quite startled me at first. The hills were covered with low bushes, heaths, sundews, geraniums, gladioluses, lobelias, salvias, babianas, and other bulbs, daisies growing into trees, purple broom, polygalas, tritomas, and crimson velvet hyobanche.

Many friends collected for me, and two baths stood in my painting-room full of wonders. The difficulty was to make up my mind what to do first. It was impossible to paint fast enough, but we can all work hard at what we like best.

Miss Duckett, who managed the great old Dutch farm of Groote Port in the absence of her brother in England, wrote and invited me to come and see her, and on the 7th of September I went by rail to Malmesbury. At the end of the line I found a covered country cart with four horses waiting and a Boer to drive them, who only talked bad Dutch, and never took his pipe out of his mouth; but he was most willing to laugh at anything or nothing. A clever little boy of ten helped him, stopping him occasionally to run after and knock down young birds, which he put into the man's big pockets alive. He said he was going to put them in a cage.

A Vision of Eden

Ostrich farming at Groote Port, South Africa. The ostriches were stripped of their feathers twice a year and apparently showed no ill effects of this persistent treatment.

We only passed two isolated farms all the twenty miles' drive to Groote Port, a most comfortable old place, with round gable ends, and double flight of steps leading to the upper floor, where the living-rooms were. Miss D. had meant me to come a week later, as eight ladies were already staying on a visit, but they all said they did not mind it, and gave me their best room. She was a regular Queen Bess or Boadicea for ruling men, and had no small work to do on that farm. Every morning she gave out over 100 rations of bread, meat, spirit, etc. Every morning a sheep was killed, and every week a bullock.

There were fifty ostriches stalking about. The ostriches are most attached couples, and seldom marry again if one of the pair dies. The cock and hen take turns at sitting on the eggs, but many wild eggs are brought in. These are hatched in an incubator in cotton-wool soaked in boiling water, and kept at over 100 degrees of warmth. They take about forty days' hatching. I was lucky enough to be in time to see the gradual entrance into life of the birds.

One day we all went up to the top of the hill behind the house, 1000 feet of steep climb, where we got a lovely view of the distant sea, and Table Mountain beyond. I found at the top plenty of small starry flowers of various colours, and large yellow thistles, many lovely bulbs, including the pink and white *Hypoestes stellata*, with eyes as changeable as the peacocks' feathers. One could not tell if they were blue or green, far less paint them. We cooked kabobs, rice, and cakes, and drank strong coffee and wine, close to a muddy pool, which they called a spring. We waited there till the sun got low again. A willow-tree hung over the pool, covered with hanging nests.

At last the smoking Boer and his horrid boy came back for me, and stopped to rob nests as they did before, cramming the poor half-fledged birds into their pockets, then rolling or sitting on them and taking them out to see if their legs were broken. They demanded pay too, before I got out of the cart, like true Boers. The rail took me back as far as Stellenbosch, a fine old Dutch town in the midst of gardens hedged with pink and white roses, beautifully tidy, and surrounded by finely formed mountains. The boarding-house in which I had a room, kept by a very old Dutch pair, was beautifully clean. After the fleas and muddle of the great farm it was a real rest, and I found the place so pretty during my walk before breakfast that I thought of staying a week, but changed my mind when the kind neighbours came to call on me, which they did in abundance, time being not much thought of in Stellenbosch.

Education is the rage at Stellenbosch, and they had insisted on placing the station a mile off, for fear of disturbing the students! They learnt psalm-singing, if nothing else, and one heard it droning on in every direction (beginning soon after daylight). It was most doleful, and the key dropped full four tones before its last verse was over. I could not help thinking of the Rev. H.J., who warned me before starting that "Africa was a most untidy country, and I should find missionaries littered all over the place." I breathed more easily when I escaped from that Calvinistic settlement, and the flowers seemed to get brighter and more abundant as I approached Table Mountain. On arriving at Mrs. B.'s I found her husband and son-in-law had just arrived before me, and the latter at once arranged to take me up the mountain. It was perfect weather for walking, the sun hot and bright, but quite cold for sitting still, and I longed for a fur-coat whilst painting in the house. A cart took us as far as the pass leading to Simon's Bay, between two spurs of the mountain topped with masses of the silver tree, whose golden flower-balls glittered in the morning light. Protea bushes and other low scrub grew near them, and the distant views of sea and hills were most exquisite, as we reached the shoulder of the mountain. After that we went up tolerably

straight, finding a succession of lovely flowers: scarlet heath, white with brown centre, a kind of broom, polygala, lovely sundews, and many rush-like plants, tough and good to haul myself up by. Grand crags and rocks surrounded us as we reached the first great flat, then more rocks to climb, and the second flat was reached at about 3000 feet above the sea. In the middle was the series of springs or streams, in which the *Disa grandiflora* delights to dip its roots. It forms a thick edging to the water.

When I was there the plants were small and green, but flowerless. Near them were tall reddish heaths and blue broom, tall masses of *Todea africana*, asphodels, tritomas, and watsonias, as well as tiny crimson rosettes of drosera and

The glory of Table Mountain, Cape of Good Hope, South Africa – *Disa uniflora* (= *Disa grandiflora*) which grows along the streams on the top of Table Mountain. The blue flower belongs to *Disa graminifolia* and behind is a fern, *Todea barbara*.

utricularia, not bigger than pins' heads, in the black bog-earth near the stream. We climbed still higher, but the clouds came down, and it was imprudent to go further where we might lose ourselves and could see nothing; so we hunted among the rocks under the clouds, and found the great *Protea cynaroides*, which has the biggest flower-head of all, larger than flowers of *Magnolia grandiflora*. It is somewhat like it in shape, only pink. The bush is the lowest of its tribe, only a couple of feet above the ground. Just then it had only its slender crimson buds on it, and those buds Mr. G. visited two successive months, but found them no bigger.

On the 10th of October I went by rail to Tulbagh. There had been heavy rains, and the country was fresh and green again. The mountains round the broad valley are very grand, with perpendicular precipices and the boldest outlines. The ground near the river was covered with bulrushes, watsonias, and arums, cultivated in the same way as at Stellenbosch. My window opened on a verandah shaded by large gum-trees, loaded with hanging nests made by a yellow finch which was perpetually building, chattering, and playing. The much-talked-of flowers were a disappointment. I was told it was a bad year, and that they were over. I walked two miles over the dry flats to Dr. Balm's mission, and as it was only seven o'clock I went on towards the hills behind, but was seen and captured, and after some breakfast was given in charge to two wild converts, and taken to the Tigers' Kluft, where the skeletons of sheep and cows were even then occasionally found. I had a hard scramble, but only found a few bulbs, ixias, and babianas, and a tiny pink pelargonium. Then Miss Balm sent me back in her nice cart with two fine horses to pull it. I had one or two other hunts in different directions, and my quarters were so pleasant to work in that I stayed a fortnight.

I paid six shillings a day at Tulbagh for my board and lodging; this included wine. An hour's more rail took me to the foot of the Mitchell's Pass, and a post-cart with four horses dragged me up it to Ceres, through the wildest scenery. Strangely twisted rocks were piled one on another, and I fancied, if one were moved, the whole would come down. Beautiful plants were growing amongst them, the most conspicuous being a purple-leaved cabbage with a huge yellow

flower like sea-kale, crassulas and blue watsonias, and bushes covered with a yellow pea-flower set in green collars, which the natives make into tea.

We descended a few hundred feet into the plain of Ceres, a filled-up lake, surrounded by mountain-tops covered with fresh snow. The plain is twelve miles across, with several villages. The little town of Ceres, buried in its trees and gardens, looked like an island in the midst of the yellow flat, which was yellow with corn, patches of leucadendron, and burnt-up grass. It was a gay little town, full of shops and people, with delicious running water everywhere – a great contrast to the place I had left. The boarding-house I stayed at was Dutch, and very dirty, unlike that at Tulbagh; but its situation at the end of the town was most delightful, as I could wander out on the flats without my hat, and find many flowers too. The situation being so much higher, the season was later.

I was told it was not safe to wander in the pass alone, because of the "black people." I thought they meant people, but they meant baboons! My neighbours at Ceres were most kind. Mrs. C. had a nice open carriage, and gave me some delicious drives. Her husband brought me several plants from the Karroo; among others a euphorbia called the milk-plant, with fleshy green fingers and tiny yellow flowers on their ends. Its roots spread for yards, with cross-bars from which rose a parasitic flower called *Hydnora africana*, like a purple star, standing alone two or three inches above the ground. Its seed-vessel swells and swells and sinks underground, till the flower tumbles off leaving a ragged ring like the crown of a pomegranate, and the fruit is as big. The natives dig it up to roast and eat; it then tastes like the flesh of a cocoa-nut.

It became very cold at Ceres; the hills were white with snow. I saw the light through the chinks in my walls and ceiling, and the broken window gave me neuralgia. In the dull weather, too, the verandah made my room terribly dark for painting. I pined for home and resolved to give up going to Karroo Gate and return to Wynberg where I found much illness in my friend's

View of the Valley of Ceres, from Mitchell's Pass, with a Cabbage Plant (*Othonna amplexicaulis*) in the foreground.

202

household, and thought my room would be valuable, so after two days started again by rail over the same road I had passed before. But towards the end of the day I reached new ground at Worcester, an old Dutch town built on an exact square of ground, the green grass edging leaving off and the desert beginning in a perfectly straight line on each side, as if cut with a ruler and knife. I half repented of not having stopped to explore its monotonous old streets, still more so when we began to ascend by a crack in the hills a steep winding pass like the Brenner, serpentining in and out of side-valleys.

I slept at a railway-hotel at Montague Road, built by the Cape Government. The next day's journey was entirely over the Karroo, covered with dusty shrubs not a foot high, with a few tiny flowers amongst them. After rain it becomes lovely and green (they say). Beaufort West was the end of the railway, where I found a lively little town and a good hotel. I had a letter from Mr. Garcia, the magistrate, and expected to see some small Spaniard with a guitar, but was much relieved at the sight of a burly Englishman. He told me "he had been fortunate enough to secure an Englishman to drive me, – the son of a late clergyman, Mr. F., – as those black fellows always drink." Soon afterwards I was accosted in a shop by the most disreputable-looking, bloated, blear-eyed old blackguard, smelling awfully of gin, and he told me he was the "gentleman" who was to drive me. I quite shied away, and when I confided my horror of him to another acquaintance there, he said "Every one here drinks, but F. and a few others are not *always* drunk!" The next morning he was an hour late, and during the first stage before breakfast he took four doses of pure spirits, half a teacupful each time, adding a few drops of water with great ceremony to show how careful he was! He never seemed the worse! He emptied five bottles during the two days he drove me, telling me long stories about himself, and all to prove what a fine fellow Alex. F. was, talking always of himself in the third person, as if he were

A Giant Protea (*Protea cynaroides*) and a Lory or Touraco painted at Cadles, South Africa. The red colouring of the bird is extraordinary since it will wash out with soap and water.

some one else. I had to pay him £15 for those two days and back-fare. We went sixty miles a day with two horses over a good road, but it was all desert, covered with small scrub, strange dwarf euphorbias, and miniature plants, which after a day's rain were said to be covered with curious flowers and green leaves, as if by magic.

A long day's rail then took me to Port Elizabeth, through a perfectly new world of vegetation in Kafirland. Every now and then there was a clearing in which was a group of beehive-shaped huts covered with skins and bits of cloth dyed of a rich red colour, or topped with black burnt bush. Near them stalked the grandest figures in red drapery and feathers, like stage Mephisto-pheles, with women dressed to match, their arms and legs covered with metal rings. I had not expected to see such genuine savages so near civilisation. They seemed too good to be real. The women carried the children sitting astride on humps which they possessed naturally, and which ladies of Europe imitate artificially, without the excuse of their being useful as they are among the Kafirs. The children were also tied on with deep-red cloth. All this rich colour was produced by rubbing with a lump of rough iron.

At last we descended from this wild country, and came to cultivation again, and to a river blue with large nymphaeas standing well out of the water, their great saucer-leaves floating round them; miles of sandy flats, with low heaths and pelargoniums of many colours, and lovely cotyledons in quantities, three or four bunches of their exquisite coral bells from one crimson stalk, sometimes salmon-tinted, sometimes bronze-coloured; but the latter grew on a larger bush. Port Elizabeth is far more like a capital than Cape Town: it is full of life and work, very clean and neat, with an excellent hotel called after my father's old leader, Lord Palmerston. Every one who could afford it lived on the hill above the town. Mr. H. called on me the evening I arrived – a most interesting man, and a great botanist. He brought two exquisite ground-orchids in his hand, which gave me work for the two next days; but I managed to get up to his house at seven the next morning, where I found him hard at work in his shirt sleeves, watering his plants, and had a delightful hour with him in his garden and the neighbouring Botanic Gardens, which he had greatly helped to make. He seemed to know every flower of South Africa, and all about them, though he passed his days in his shop selling tea and sugar. He told me I was only wasting my time in the town, and ought to go at once to "Cadles"; so I did, driving over twenty-five miles across the "flats" to it.

Cadles was perfect quarters: a sort of farmhouse rather than a hotel, with the kindest of hosts and hostesses – the latter almost immovable from dropsy. She had half a dozen nieces of adopted daughters who did the work, looked after everything, and kept the place lively and in order. One felt more like a friend than a boarder. I had an upstairs room, opening on a verandah with a window at the other end, looking over the farm and offices across the kluft to the distant mountains to which it leads, and from whence the delicious water is brought to Port Elizabeth. All round the house is a sort of hilly tableland, covered with coarse grass, coloured with pink and white watsonias and other flowers. This tableland is cut up by great klufts or cracks, loaded with rich forest-trees, and tangled with creepers, wild vines with brown stalks and tendrils, plumbago, ivy-pelargonium, and various gourds: one they called "wild Chili" (*Cephalandra palmata*), with hanging egg-like fruits of scarlet or green, made the most exquisite festoons, but was poisonous. There were great candelabra-like euphorbias fifty feet high, and the calodendron or wild chestnut covered with large bunches of lilac flowers.

I had been long seeking a good specimen of the *Protea cynaroides*, the biggest of all its race. I saw the plant on the Cape Table Mountain, but searched in vain for a flower. I had mentioned this to a friend I was lunching with at Port Elizabeth. He rubbed his head for a moment, then said: "I know

A Gifbol (*Boöphone disticha*) at Grahamstown, South Africa. Hanging above is a *Zygophyllum* and the trailing plant with reddish, daisy-like flowers is a *Mesembryanthemum*. To the left of the big bulb is the curious brown-flowered Toad Cactus (*Stapelia variegata*) and in the left corner a pink orchid probably *Satyrium longicolle*. The butterflies are *Hypolimnas misippus*.

where it is!" rushed into a neighbour's house a few doors off, and brought me out a magnificent flower. I almost cried with joy at getting it at last, I had missed it so often; took it with me to Cadles, and painted it there. The bracts were like pink satin, tinted at the base with green, and a perfect pyramid of yellow flowers rose in the centre. While I was eagerly at work over this gorgeous flower, my landlord brought me a lory or touraco, with a lovely red beak and eyelid, and its green wings lined with that deep magenta colour which has made this bird famous; for it washes out in soap and water, and, what is still more strange, the bird is said to have the power of recolouring it. The colour when washed out has been analysed and found to contain much copper. I was dumbfoundered when the bird was brought me to paint, as I could not give up the protea, so made a compromise, and managed to show only its head and part of the famous wing, the rest of the bird being hidden by the flower and leaves.

It was hard work to paint all the beautiful things they brought me at Cadles. On Saturday Mr. H. telegraphed to know if I were still there, and came over and spent his Sunday in taking me to the head of Van Staaden's Gorge. They gave me a perfect pony from Basutoland, a strong roan, which treated me as if I were no weight at all, and both walked and cantered to perfection. We soon reached the hills, and the aqueduct leading one way to the gorge and the other to Port Elizabeth, whose streets are bordered by delicious running water, and every garden has its fountain. The gorge was very narrow, and bordered by reddish cliffs, through which ran the clearest of rivers, with deep pools amid masses of ferns, pelargoniums, watsonias, blue hyacinths, yellow and white daisy-trees, everlastings, polygalas, and tall heaths. Under them we found precious parasites (harveya), white, pink, and scarlet, while above, in rocky cracks, hung euphorbias and zamias, or Kafir bread-trees. At the head of the gorge we came to a waterfall and a reservoir.

Just where the water bubbled out purest and freshest were quantities of a small pink and white disa and lovely droseras. We returned over the windy downs on the other side of the hills amidst acres of protea bushes of different sorts, and huge everlasting plants standing a yard or two above the ground, with white velvety leaves round a thick stalk, surmounted by a cauliflower head of white petals and yellow stamens. These looked like tombstones at a distance. A gentleman who had been also to the gorge cut one of the great things down, and carried it home over his shoulder for me to paint, for which I was grateful, for it was no small weight.

Mr. H. often came over to Cadles, and his enthusiasm as a flower-hunter was quite equalled by his genial good-nature. At last I made up my mind to tear myself away from that charming place, and found it not so easy to accomplish. I missed one chance of a return carriage through my deafness, then had to share one with people who could not get up in the morning. I returned to Port Elizabeth, and then the railway took me through the Adda Bush: a flat, swampy locality, full of spekboom trees, which are said to tempt the elephants down close to civilisation, and herds of them are still found there. Some twenty miles of dense tangle prevents mankind from interfering with the poor beasts, and the climate suits them better than it does their enemies. It was a great relief to begin ascending and the road became very steep and wonderfully made. At the top of the pass we came to groups of the oldenburghia, a most striking shrub which grows only on these hills, and on the very tops of them. The hills became very bare as we descended to Grahamstown, which is built in a perfect hole; but it was pretty inside, with many nice people in comfortable houses with lovely gardens. Its Botanical Garden was particularly well kept, full of interesting plants, the wild hillside and rocks coming down into it at the back. Mr. H. had given me an introduction to Mrs. G., the wife of a watchmaker there who had built a curious house with a tall tower, and had seven sons all fond of natural collections of different sorts. These young men brought me constant relays of beautiful flowers. I stayed in a quiet room at the railway-hotel, where I could work well with a good light, and seldom went out except for drives with Mrs. A., a wonderful old lady of over eighty. Her husband, Dr. A., was the one man of the place, and full of information. His garden, which he planted forty years before, was well worth coming across the world to see, and full of strange plants, mostly native. The *Strelitzia augusta* was in full flower, as were many varieties of aloes. The biggest of all aloes had been called after his

South Africa

View from Dr. Becker's verandah over the Kowie River, Port Alfred, South Africa. The tree is *Tecoma mackenii* and has a Cape Weaver (*Ploceus capensis*) perched upon it.

relative, Mrs. Baines. I had heard her name ever since I entered South Africa as the great authority on all sorts of natural history, and was delighted when she walked into my room one day and said she had come from the country on purpose to see me.

She showed me some of her own paintings, stippled on white paper, with a line of neutral tint round the edges to raise them (done much in the way old Anne North did her flowers in the year I was born). She had painted many of the stapelias, and brought me two to do – brown and yellow stars with a most evil smell. They attract flies, which try to get at the nectar, and thereby fertilise the flowers, then are rewarded by being themselves caught by their legs, and probably have their lives sucked out of them. Their efforts were frantic and most futile to get free from the tormentors.

On the 6th of January 1883 I started (with Mrs. B.) in a private carriage for a six hours' jolt-jolt to Port Alfred. We crossed low hills covered either with dwarf scrub and euphorbias, or as bare as the Sussex Downs, Kafir huts occasionally on their tops, with patches of Indian corn and scrub in the hollows. Dr. Becker had come to meet me, meaning me to stay at his nice place up the river with his wife; but as usual a companion (when I do have one) spoils all those nice arrangements, and he could not find room for two. Even when we reached the hotel there was no way of putting us both up, except by sharing a bed or sleeping, one of us, on two chairs (which my friend did). The situation of the hotel was pretty, with thick scrub round it on three sides, and a view over it to the white sandhills at the edge of the sea.

Ten minutes' walk through the woods or over the marsh below took us to the sandhills, where different people were camping with their waggons during the summer months for sea-bathing. At the end of the river a pier was being slowly made by convicts, and a railway to meet it from Grahamstown will in course of time make the

pretty place a cheaper and easier resort for its dried-up inhabitants. Mrs. T., the wife of its engineer, took us in her boat a few miles up the river, where I drew some old zamias and strelitzias hanging over the water. Then we went to see Dr. Becker, whose house stood high over the great Kowie river just at its bend, with lovely views both ways. He had had the good taste to leave the giant euphorbia and other native trees. He had also a collection of stapelias and other small prickly plants; some of them were almost invisible without a magnifying glass, but most interesting.

Mrs. B. only stayed two days, but her sister and her husband, a brother of Dr. A.'s, came and drove me back in their carriage, which I much enjoyed. We made two days' journey of the twenty-eight miles, and stopped every five minutes. The whole way was pretty – one part especially, where there was a kluft full of wild date-trees bordering the stream and rocky banks; the place bristled with aloes and euphorbias. A kind of asphodel was growing up from the ground like a spear stuck in by its handle-end, and no leaves.

Dr. A. had asked me to stay in his house on my return, but all his rooms were so darkened with vegetation that I despaired of finding one I could see to paint in. At last I was taken across a sort of drawbridge to an attic or wide verandah, dark with a solid mass of creepers, bougainvillea,

banksia-roses, etc. It was full of old boxes of rubbish, and some crazy steps led into another attic with a small room at either end, one of them given up entirely to a colony of bees, which had built in a corner and resisted intruders; the other belonged to a son, who was on the opposite side of the colony. There I settled myself, digging out a hole in the bougainvilleas, which let in a blaze of rosy light through their flowery wreaths. It took me some time to make the opening sufficiently large to get natural-coloured light enough to paint by. One side of the room was separated from the verandah by broken green-house-sashes, one of them off its hinges. The roof itself was all of a slant, books, hats, pipes, and various treasures of men in delightful disorder, but it was deliciously quiet and out of the way. I bolted myself in at night, and shared the whole storey with the bees and an occasional rat, bird, or lizard, and through my hole in the bougainvillea I looked on wonderful groups of aloes of many species, and other rare things collected by the Doctor in his long African life.

I had determined to go overland to St. John's, and the Governor, to whom I wrote, kindly telegraphed to the different magistrates to be good to me if I came in their way. One friend in Grahamstown told me it was dangerous, and some people had lately been washed away when trying to cross the rivers; but the doctor said it was not necessary to cross them at flood-time, and I might as well go. He went to Cape Town to his Parliamentary duties, and the sunshine went with him. My bower was soaked with wet. There was a perpetual succession of dawdling, dull society, and I fled down to pack at Port Elizabeth, whence I sent my paintings home. I slept at Coerney in the Adda Bush, where there was a nice little hotel, full of children out for the holidays. The plants seemed all fleshy or thorny, but there were some pretty bulbs, one of them flowering pure white in the morning, and changing through pink to crimson at night. Mrs. A. sent for me to sleep again on my way through Grahamstown. The post-cart was to start at eight.

My next stop was at Beaufort which we reached at four. The weekly cart started at five the next morning, so I set out without even getting a cup of coffee. The road along the Kat River was very pretty; but Balfour, where I meant to stop, looked

Flowers of St. John's. Beginning on the right at the top, there is a dark blue *Coleotrype natalensis*, a purplish red balsam (*Impatiens* sp.), clusters of the small white flowers of the White Pear (*Apodytes dimidiata*), the red and black seed-vessels of which lie at the foot of the basket on the right, and a *Loranthus* with reddish flowers. In the centre is a white *Pavetta*, with immediately over it the purple-red flowers of *Vigna vexillata* and a solitary flower of the yellow *Sphedammocarpus pruriens*. Returning to the right are two pale orange flowers and narrow-tendrilled leaves of *Littonia modesta*, a single lily, followed by the more showy deeper orange *Crocosmia aurea*, a large *Hibiscus* with purple centre, the purplish *Grewia lasiocarpa*, with an *Ipomoea* below. Lying in front is a branch of *Acridocarpus natalitius*.

dressy, and in a hole, so I went higher and was left at "Therous."

The air was delicious, but it was like a dozen low alps I know in Switzerland: bare alps, with no shade except in the deep klufts. Even there the trees seemed round-headed and commonplace, with not even euphorbias to mark the continent. It was a solitary house kept by pious temperance people, very clean, and the food particularly good; but I found no new flowers to tempt me to linger a week, and felt I should go mad before it was over. The flowery season was past. I toiled twice to the top of the pass and back in vain, to find nothing worth painting. The carriage came at last, and four bullocks dragged me up the hills. The views were certainly fine, but, as Mr. Lear wrote, "Cumberland is nearer and prettier," and there was nothing in it characteristic of Africa. The road was very bad, but I had a capital Kafir to drive me, and before the after-glow had faded out of the sky we had got over all the difficulties, down to the plain, and reached a decent inn before dark. The next morning I reached Queenstown by breakfast-time. It was a dreary country, and I was glad to escape by train to King William's Town. I had telegraphed to a friend who met me on arriving. In the Botanical Garden we saw the poor director. They settled that I must go to see the Perie forest, and stay with a family there, whose father was a sawyer, and had a second mill at East London, while his son managed the one in the forest. It rained all night, but my friend was clear to start, so off we went, he riding, I driving a country cart over the most awful roads. We lost our way, and went short cuts. Once when it seemed too bad for wheels to pass, I got out and fell in the mud; after that I slipped on some stepping-stones and got soaked in the river from head to foot, in which state I arrived at the house. I had to change everything at once. The good Frau was very hospitable, and gave me her parlour, making me up a bed with a small sofa and a chair; and there I stayed four days, for the rains came on with a vengeance. Mr. C. got home next day somehow, and sent a carriage back for me. The driver said I must not wait a minute or the river would be impassable. Mrs. S. said, "Don't go, it is dangerous," and sent the man off, who found that it was! He had to invent a new way of getting home over the tops of the hills, so I settled myself to paint Kafir-plums in the "parlour."

At last the rain ceased, the sun shone, and the good Frau took me and three children three miles through the forest in her ox-waggon to the saw-mill. I enjoyed it much, and felt far safer in that long wood-waggon than in a cart with horses. We stopped near the saw-mill, and stayed all day in the cool forest, where I got sketches of the great yellow-wood trees, covered with lianes and other creepers. The Kafir boom, or large erythrina, was a tall tree, with great thorns or knobs over its stems. As it was again fine, and no cart appeared, we stopped a wood-waggon and engaged it to take me and one of the girls who was returning to school to King.

We jolted merrily along with six fine bullocks, sitting on the floor of the long springless cart, till we reached the top of the hill, and saw the promised carriage going on rapidly by another road, and Mr. C. jogging after it. They turned and looked at us, but went on; so did we, and we were not overtaken till we reached the half-way house. The cart was driven by the same man who had come after me in the storm, and I soon found out he was the famous driver, Mr. Hewitson, of whom I had heard so much. He showed me the height the river had risen in the late rains, and that it would have been certain death to have attempted to cross the river then. We picked up a nice young Englishwoman and her child, who were going with him to Mutata to join her husband, and he persuaded me to go at the same time and on to St. John's.

The scenery got more and more magnificent as we neared St. John's River, and came down a "good road," which seemed to me much like the bed of the stream. Just as we reached it, we met four gentlemen riding, who stopped us; and Mr. O., the Resident said I must come at once to his house at the mouth of the river; so I was put into a boat, and said good-bye to Mr. H. I enjoyed the grand scene, all glittering in the sunset lights. As it got dark the great gates on either side the river looked prodigious, with their rocky crowns, three hundred feet of sheer granite precipice a thousand feet or more above the water, the intervening slope being covered with unbroken forest down to the river's edge; only on one side was it practicable to walk, and there much blasting had been required to make a tolerable riding road.

South Africa

At the end between the two cliffs was the sea, with its double line of breakers over the bar, which only permitted vessels to cross at the very highest tides. My host had ridden on before to warn his wife of the invasion.

It was very peaceful to wake the next morning and to be in no hurry. They were late people at the Residence, and I had a walk all over the place before I saw any signs of breakfast. The resident magistrate was a curious mixture. He had tried many trades: sailor, doctor, lawyer, and parson. I suspect the bishop's adopted daughter had something to do with these changes. She nursed him through a fever, he married her, and all his dreams of converting the heathen departed. He was now administrator of Pondoland, and had strong convictions that white and black could not live together, and that the latter must go sooner or later, if the other comes. He had plenty to do, for the different tribes were always at war, and he was constantly riding off for many days at a time, trying to patch up peace among them. St. John's settlement was only three years old, and the growth of the trees and fruits planted there since was something extraordinary; but it was a damp, hot climate between those hills, and the natives had the good sense to avoid it, and to live on the high table-tops above.

I waited in vain for the chance of going on to Natal overland; it always got more and more improbable: 120 miles through a country where the tribes were actually fighting. So I painted more flowers. A loranthus (tree-parasite, called by Anglo-natives "honeysuckle") was very interesting. When I touched the end of the bud, if ripe, it suddenly burst, and the petals sprang backwards, while the pistil in the middle, which was curved like the spring of a watch, was jerked out a yard or more. A tree called the "wild pear" was just then in full bloom, and very conspicuous in the landscape, with its fine white flowers.

The forests were so full of grass-ticks that, after no little suffering from them, I generally walked on the golden sands at the edge of the sea by preference. They were covered by a perfect network of *Ipomoea pes-caprae*, with millions of seeds and lovely pink flowers, the long trailing branches often twelve to twenty feet in length, radiating from the central root like a huge green octopus. There was also a handsome pink-flowered bean spreading itself in the same fashion on the sand. At the end of the sands the cliffs rose 300 feet, with the signal-station at the top, and views over grassy downs, far away down the coast as well as back to the river, with its great gates 1300 feet high. Below that hill I found beds of dwarf hibiscus, with a white flower and claret-coloured eye, and masses of blue agapanthus and tritoma.

Bishop Callaway arrived while I was there. Mrs. O. was his adopted daughter. He told me many interesting stories of his African life. Once a snake of eight feet in length was found coiled away in his study, and one of the men said he had seen it climbing up behind the books three days before, and though he was shaking with fright, he did not say anything, as he supposed it was one of Dr. Callaway's friends who was too big to put in a bottle. They believed in transmigration of souls there. The bishop also told me several stories of people who had been partially paralysed by a very poisonous snake, the tuamba, which merely raised its head and looked at them, then glided away satisfied. Both he and Mr. O. said they had known people who had lost all power for two or three days afterwards, and for the time were unable to move from the spot.

There were plenty of wild bees in the woods, and when the natives wanted honey they put a stick into the hole first; if the bees came out, they knew all was safe, if not, they knew a snake lived also there. The bees were accustomed to him, and took the stick for another snake. Once we spent a day on the other side of the river, close to the sea's edge. The rocks on that side went down to the waves, with lovely pools and aquariums amongst them, and were fringed above with *Strelitzia augusta* and aloe-trees, besides various fleshy plants, mesembryanthemum and portulaca. It was by far the most beautiful side, but, though so near, very difficult to get at. There was a spring a little higher up, to which we used to send for drinking-water.

When at last it was decided that the overland journey was impossible, and that I must go by sea, the steamer did not come! and I thought I never should get away, but another fortnight brought a large steamer, the *Lady Wood*, often talked of but never seen before at St. John's. On the 11th of April she took me over the bar easily, all the population

213

Above: *Strelitzia alba* (= *Strelitzia augusta*) at St. John's Kaffraria, with trees of the same in the background and Cape Honeysuckle (*Tecomaria capensis*) trailing over the vegetation on the left.

Opposite: A remnant of the past near Verulam, Natal, South Africa. This trio of old aloes (*Aloe bainesii*) were the only specimens to be found in the region when the paintings were made.

collecting on the shore to watch the feat. At nine the next morning we entered the harbour of Durban, and a good-natured young man brought my things and myself by the tram to the Marine Hotel there, where I found a nice airy room.

In the evening Mr. A., Donald Currie's agent, drove me up the hill to dine and sleep at his house, as his wife was going away the next day. They lived close to the Botanical Gardens. Their house was quite hidden by the bright blue ipomoea, generally called "morning glory." I never saw it more lovely. The hedges all over that hill were hung with other ipomoeas, bignonias, tecomas, thunbergias, and a lovely white creeper from Barbados. I saw also an exquisite hibiscus growing tall, like a hollyhock, with a deep blue eye. In the gardens I found splendid zamias of all sorts, and stangerias which came from St. John's. Their leaves are so like large fern-fronds that they have deceived botanists, as they did me at first.

Then I took the coast railway due north to Verulam, passing five round-topped euphorbia trees, and then rich crops of sugar and corn,

cultivated by most picturesque coolies from India, all over bangles and gay garments. After that, I drove seven miles farther through pretty country up to Tongate, to stay with Mr. and Mrs. S. Their house had never been finished, but was already rendered dangerous by white ants, which were eating all of it that the damp had left. It stood on the top of a hill with a lovely garden, and distant views all round. Cotton with pink, white, and yellow flowers; sugar, coffee, and fruit-trees were there in quantities, all in good order. The cotton-harvest was going on, and a curious crowd of Zulu men and women were at work over it.

A most enthusiastic botanist came to see me, and said I must go with him to see a group of aloe trees forty feet high, the only ones left in that country; so I went over to Verulam to stay a couple of nights, and he drove me over. He was a butcher by profession, but had bought considerable property, and started a large sugar-mill. Near this we found the noble group of trees on the bank of the river. The trunks measured two feet through at a yard above the ground, and rose to perhaps twenty feet of stately gray stem, then split into forks, which re-split into numberless pairs of great leaf-bunches, bearing three to five spikes of scarlet flowers, like red-hot pokers, in July, when they might be seen forty miles off. They were the sole remains of forests which had disappeared in that part of Africa, perhaps for centuries, and even those three trees have been cut down since I was there; so I have been told. Mr. H. sat and watched

me at work, much pleased to see his dear aloes at last done justice to. He said not even Mrs. S. had been to see them, and when he wrote a description of them to Kew, they had coolly asked him to cut one down and send them a "section" for the museum!

I found there was no way of getting either to Zanzibar or Mauritius for six weeks, when a ship was going to the latter place, touching at Madagascar; but I was ill and home-sick, and decided to go home and take a rest, so left in the *Melrose* on the 22nd of May; had another talk with my botanical friend at Port Elizabeth, two days with the Gambles in their new house among the silver-trees at Wynberg and reached home on the 17th of June 1883. The three months at home were delightful, and gave me fresh strength and courage for the task I had still set myself to do.

Mr. Fergusson had already planned another room at Kew, promising that it should be finished by the time I returned from my next expedition, and when I left England on the 24th of September the walls were already up. Meanwhile I enjoyed seeing old and new friends again, and reading the books I had longed for when out of their reach in Africa.

Before starting on another journey, I went out to see my sister and her family at Davos, stopping on the way to pay a long-promised visit to my old friend Madame Sainton, who had bought an old country house seven miles from Boulogne. Davos was too cold for me and I soon came home again.

CHAPTER XII

Seychelles Islands
1883

On the 27th of September 1883 I left Marseilles in the *Natal* and on the 13th of October I landed in Mahé at daybreak. The lovely bay is surrounded by islands, Mahé encircling two sides of it, the mountains rising nearly 3000 feet above it. The cocoa-nuts mounted higher than I ever saw them do before, rich green fruit and spice trees linking them to the natural forests above. On the top of all are fine granite cliffs, not in the long lines of St. John's, but broken up and scattered amongst the green. The little town with its trees and gardens is squeezed into a narrow valley, so that only the houses along the sea's edge were seen as we approached it over the long pier. On each side was a wide stretch of black mud and sand, covered with exquisite turquoise crabs with red legs, so beautiful that I dropped my bag and screamed with wonder at them, to the amazement of my porters, who said coolly they were not good to eat! An avenue of tall sangu, dragon, and melia trees bordered the road up to Government House, a pretty low bungalow. Hanging baskets made of fern-stems ornamented the verandah, and a tolerable specimen of the "coco de mer" with its first nut upon it, was planted in front.

Mr. and Mrs. B. received me most kindly, making me feel at home, and I walked across the low pass to the other side of the island that same afternoon. It was exceedingly beautiful. Nutmegs, cinnamon, and cloves were all growing luxuriantly, but the people are too lazy even to pick them, and I crushed under my feet the purple fruit of the latter, which had been allowed to go to seed and fall from the trees, no one troubling to collect

the buds at their proper season. Every little hollow was filled with vanilla gardens. That plant being trained on espaliers like old-fashioned apple trees, the lazy people could attend more easily to its culture, and they soon realised a good return for the money spent upon it.

Walking was said to be impossible, so my two first expeditions were made with four bearers and a chair. They charged fifteen shillings for the day; but as I found that I walked on my own feet most of the time, and the men were greatly in my way, I never had them again.

Many people kept tortoises as pets. It was considered a rich kind of thing to do in the Seychelles, as keeping deer is in England. One family I heard of always named a young tortoise at the same time as a new baby, keeping it till the child grew up and married, when it was killed and eaten at the wedding feast. In one place I saw about thirty, which were twenty-two years old, and about two feet in length. They grew fast up to that time, then seemed to stand still for centuries. Like the double cocoa-nut trees, they were all descended from those brought from the island of Aldabra, where alone the great tortoise is still wild.

On the 22nd of October I started in the little Government sailing boat for Praslin, and after three or four hours without wind got slowly over the twenty-five miles of sea, and landed at midnight in the lovely moonlight in Dr. H.'s little boat, two hospitable dogs licking my face and hands as I jumped on the sands, and nearly knocking me down. Johnnie B., a nice boy of thirteen, had been sent over to take care of me; and we had a real good time in Praslin with the H.s in their nice new house. Half an hour's row across took one to the island of Curieuse, the only other island where the "coco de mer" grows wild. To see and paint that was the great object of my visit; so the next morning after my arrival Mr. H. took me round the east side of Praslin in his boat. We passed close along the shore among beautiful boulders of salmon-coloured granite, grooved and split into fantastic shapes by heat or ice. Many of the little islands had waving casuarinas on their tops, while bright green large-leaved "jakamaka" bordered the sands of Praslin, varied by patches of cocoa-nut and breadfruit. Above were the deep purple-red, stony-topped hills, with forests between, the famous coco de mer palms shining like golden stars among them. At last we ran into the valley of the coco de mer: a valley as big as old Hastings, quite filled with the huge straight stems and golden shiny stars of the giant palm: it seemed almost too good to believe that I had really reached it.

There was a thick undergrowth, and we had not started till late, so that I could only make one hasty sketch of a tree in full fruit: twenty-five full-sized nuts, and quantities of imperfect ones, like gigantic mahogany acorns. The outer shell was green and heart-shaped; only the inner shell was double, and full of white jelly, enough to fill the largest soup-tureen. The male tree grows taller than the fruit-bearing one, sometimes reaching 100 feet; its inflorescence is often a yard long. The huge fan-leaves of both trees are stiff and shiny, and of a very golden green, different from all other palms (except the cocoa-nut) in colour. We had sent the boat round and walked up the valley, in which were more than a thousand of the giant-palms; but we could not stop, and I never was able to get there again. We walked across the island to the north shore.

The island of Curieuse was only half an hour's row from our shore. One poor old Englishman lived there, having married its heiress. He lived in an old house which had never been altered for two generations, and was picturesquely dilapidated. A long double avenue of large lilies or crinums bordered the road approaching it from the shore for a good hundred yards. These flowers stand five feet high, bending their great crimson and white petals towards the light, cocoa-nuts and mangoes

arching overhead. Near the house were groups of huts, wedged between boulders, and thatched with the great fan leaves of the coco de mer, the stalks forming ornamental points at the corners, and finishing the roof into a curve like the gates of a Japanese temple.

There were many of these trees on the island of Curieuse, and a path was cut to one of the biggest, with a pile of boulders behind it, on which I climbed, and perched myself on the top, my friends building up a footstool for me from a lower rock just out of reach. I rested my painting-board on one of the great fan leaves, and drew the whole mass of fruit and buds in perfect security, though the slightest slip or cramp would have put an end both to the sketch and to me. Bright green lizards were darting about all the time, over both the subject and the sketch, making the nuts and leaves look dull by contrast.

Manioc roots and rice with some cocoa-nuts form the chief food of all people on these islands, but when they choose they can catch plenty of fish. There is an endless variety, and they seem all good food to the blacks. By scraping up the sand just above the receding tide, or even under the water, great quantities of pretty bivalve molluscs could be procured, which made an excellent soup.

The valley behind the house at Praslin was full of "latamir" palms; the most useful of all perhaps, for in those islands the houses are all thatched with them, though the roof-ridge is finished with the bigger fans of the coco de mer. Over those palms, and over the boulders, were twined long green snake-like leafless stems of the *Vanilla phalaenopsis*, the hanging ends turning up with two slender leaves just before the bunch of buds and lovely white flowers. Another beautiful orchid (*Angroecum eburneum*) grew on the hills of Mahé, and these two were the only really beautiful native flowers I saw in the islands, all the rest being more or less foreigners, chiefly from the East or West Indies and Madagascar.

Another day I went with Dr. Hoad to La Digue, a large fertile island containing six hundred people, who possessed grand cocoa-nuts and consequent riches. Many boats were anchored there, which took the oil to Mahé; but water was scarce, and the manager told me food was also. He looked half starved. We reached the island through a gap in the coral rocks, so narrow that they might have been touched on each side with our hands. In bad weather it is impossible to land, and it is never easy. The colours were marvellous on these clear seas, and that day while the doctor was paying some extra professional visit in another clear bay, I stayed still in the boat and looked down through the water at gorgeous flower-beds of coral, with blue and gold fish darting over them – things so lovely that I hardly believed them real.

We occasionally caught odd fish while waiting for the doctor. One was a pilot-fish, with a flat cohesive jaw, which stuck to the shark, and got the benefit of his fishing. Another had a long whip tail, and once they speared a hawk's-head turtle or "carré" which a shark was trying to get into his mouth: a rather large morsel, as it was over three feet long. I bought it off the sailors for £2:10s., the value of the tortoise-shell back. Catching them is the principal aim of sailors in those islands, and they divide the profits made by each boat, one man often making ten pounds in a season. The one I bought was found to be full of eggs, which were collected in a pail, buried in the sand near the house, and kept till they hatched, after which they were kept another six weeks with difficulty, as they have an inclination to run into the sea as soon as they leave their shells, and would be quickly gobbled up if they did. They are fed on fish, and some of the natives keep them till their shells become saleable; but to do this is more trouble than they are worth.

One morning I scrambled up to another valley full of coco de mer palms (far larger than any I had seen before), as well as other giant palms, pandani with stilted roots, and magnificent ferns. We had to cut our way, and I sat down and slid over slippery rocks, and dropped into foot-holes cut for me on palm-branches or anything else that came handy. In those forests there were so many thorns on the tree-stems that one had little power of holding on by one's hands, and it was no easy matter getting down an almost perpendicular bit of forest-covered hillside; but when it was done I was well rewarded.

After crossing the stream to the other side, a waterfall fell fifty feet over a wall of rock hung with ferns, into a pool among the boulders, half hidden by foam and steam. Above and all round were huge coco de mer trees, loaded with full-

grown nuts or long sweet male flowers.

It was no easy matter to get away from Praslin. Many boats went and came, but they were full of dried fish and natives (equally unpleasant at close quarters), and except during exceptional winds rowing-boats took eleven hours, sailing-boats any time. But at last Mr. B. sent me one of the Government whale-boats, and I had a most agreeable row over in eleven hours with Johnnie B., and a kind welcome at Government House; but it was not comfortable quarters for long. The place was just under the old burying-ground, close to the dirty town; so, as Mrs. B. wanted a dry room for the baby, I packed up, gave up mine at once, and went off to Mrs. E. She had turned her nice dining-room into a room for me, which was delightful, opening on the wide verandah, with the glorious view over the bay and grand

Native Vanilla (*Vanilla phalaenopsis*) hanging from a wild orange, at Praslin, Seychelles.

A Vision of Eden

Above: Female and male – the taller – Double Coconuts or Coco de Mer (*Lodoicea maldavica*) at Praslin, Seychelles, with Round Island, Felicité and a portion of La Digue in the distance.

Opposite: A selection of wild and cultivated flowers from the Seychelles with a small flowering branch and fruit of the Puzzle Nut (*Xylocarpus moluccensis* in front of the vase. On the right is a crimson Coral Plant (*Russelia equisetiformis*), then the whitish Horseradish Tree (*Moringa oleifera*) and a yellow-brown *Strophanthus*; below the latter are the pink flowers and open seed cases of a *Tecoma* and a white *Ipomoea*. In the centre are *Cerbera odollam* with on the right a Crape Myrtle (*Lagerstroemia indica*) and *Clerodendron thomsonae* in fruit.

mountains beyond, and their ever-changing clouds and shadows: great nodding scarlet hibiscus bells as foreground.

I had perfect peace. Mr. E. was at his office all day; his wife, a real sweet English lady, was my companion, with her model servant Madalina, and her two boys. Instead of drains, I had the scent of forty feet of stephanotis in flower along the rail of the verandah, mixed with *Roupellia grata* and jasmine. Cocoa-nut trees hung over and framed the different views, the island of St. Anne's making one perfect picture in itself. Seven other islands came into view from my window, as well as the lovely coast of Mahé itself, with its purple granite rocks above and the exquisitely coloured sea below.

We had the most delicious fruit in abundance. Sixteen large pines could be bought for sixpence, the mangoes were as good as any in India, and papaws, most excellent, also melons. The wild raspberries were brought down from the hills in baskets made of banana-leaves, shaped like boats; the fruit were not good raw, but slightly stewed, and then left to cool, they were delicious for the next day, when we ate them with cream. Some black people kept four cows close to us, so we always had plenty of milk and cream. Beyond the house was a path along the lovely coast for some miles. I generally went for a walk along it before breakfast, diverging up to the hills at different points. Centipedes are the only evil things in the island. They live in the hollow between the wood and the palm or pandanus stalks which line the walls of houses, and come out when the evenings are damp. They are three to five inches long, and their bite at the time is like touching a red-hot poker, but it goes off without leaving any permanent hurt. Our dog Snap always gave notice when he saw one, but took care not to get within dangerous reach of it.

At Christmas and New Year the whole population got mad drunk. All the black and brown people began by going from house to house, wishing *banana* or *bonne année*, and in return got a glass of rum, or money to buy it. At night we heard singing and raving all round; it was like the island of lunatics, and we barred all the windows well before going to bed: to sleep was impossible. Mrs. S. asked the Judge and the O.s to dinner on New Year's night. The Judge (or some one for

him) sent word that he was too drunk to come, and poor Mrs. O. said she feared it would be impossible to get her husband sober enough to walk there. The Judge afterwards had to put off the sessions for three days, because he was too drunk to hold them!

Rice and other provisions necessary to the people became so dear that they were half starving, and there seemed little chance of my getting away, and no chance of getting on to Mauritius.

I stayed three weeks with Mr. and Mrs. W. He had begun life as a blacksmith in Somersetshire, but fancied he had a "call," and came out to be cured of the idea. After five years of perfect loneliness in Mahé, they had now two fair little children of their own and sixty black ones to look after. The schools were originally intended for the children of slaves, but now that none existed others were taken: I could not make out by what rule. They all seemed very happy there, and did not puzzle their brains with too much learning. Report said they were famous thieves when they went down into the lower world, but that lower world had also a great reputation for untruth. I found them quite good-natured and honest when among them. Psalm-singing seemed their chief study; morning, noon, and night it went on, and I rejoiced in being blest with only one ear that could hear. The situation of Venns Town is one of the most magnificent in the world, and the silence of the forest around was only broken by the children's happy voices.

From that flat-topped, isolated hill, one saw a long stretch of wild mountain coast, and many islands, some 2000 feet below, across which long-tailed boatswain-birds were always flying; behind it, the highest peak of Mahé frowned down on us, often inky-black under the storm-clouds. They were gathering round it when I came up on the 7th of January, and for a whole fortnight the rains came down day and night, showing me wonderful cloud-effects, dark as slate, with the dead white capucin trees sticking through like pins in a pincushion. There were few living specimens of any age, but those were noble ones, the young leaves a foot in length, looking like green satin lined with brown velvet, and growing in terminal bunches at the ends of the woody branches. They seemed to me much like the gutta-percha trees of Borneo, but I could make out nothing certain of the flowers, and was told "it had no flower," or a "red flower," or a "white one," each statement most positive, from those who lived actually under the trees! The nuts every one knew, and collected them as curiosities. Flowers were sent afterwards to England, and Sir J. Hooker declared it a new genus, and named it *Northea seychellana*, after me.

Mr. C., the Messagerie agent, thought he might get me off by the next mail. It was nice to be at home with Mrs. E. again; but I had all the uncertainty of whether the mail would take me, and when I got to the ship's side it would not. There had been thirteen deaths from smallpox in January, so I agreed to go into quarantine in Long Island with the C.s, who also wanted to go home. Another man joined them, and for ten days all was peace.

I painted continually, and looked at the lovely views, and the marvellous colours of the sea. The first ten days on that quarantine island passed peacefully; after that some of the inmates took to playing tricks on me, and I thought they would rob and even murder me. God knows the truth. Doctors say my nerves broke down from insufficient food and overwork in such a climate. There being no banker, Coutts had sent me £200 in notes, which were stitched into my clothes; and for two days and nights, I tied up my door, barricaded my window, and was in fear of my life, hearing things said behind the low divisions, which they tell me never had been said. The ship came at last, and we got home; but the same troubles followed me till I reached England, when I was again among my friends, and able to enjoy finishing and arranging my paintings in the new room at Kew, trying to forget all that dreadful time.

Marianne North's health was breaking down, her nerves partly destroyed, but the old spirit was still there; and till she had finished the last bit of the task she had set herself, and painted on the spot the strange forest-growth of Western South America,

A view from the artist's window at Mr. Estridge's, with the harbour of Mahé, Seychelles, below. The showy shrub is *Hibiscus liliiflorus.*

she would not allow herself to rest. Just before she started on this last long journey, a great pleasure came to her in the following letter:

Osborne,
28th August 1884.

Madam – The Queen has been informed of your generous conduct in presenting to the nation, at Kew, your valuable collection of botanical paintings, in a gallery erected by yourself for the purpose of containing them.

The Queen regrets to learn from her Ministers that Her Majesty's Government have no power of recommending to the Queen any mode of publicly recognising your liberality. Her Majesty is desirous of marking in a personal manner her sense of your generosity, and in commanding me to convey the Queen's thanks to you, I am to ask your acceptance of the accompanying photograph of Her Majesty, to which the Queen has appended her signature. – I have the honour to be, Madam, your obedient Servant,

Henry F. Ponsonby.

Miss Marianne North.

CHAPTER XIII

Chili

1884

ABOUT THE MIDDLE of November 1884, I started on my last journey. All the biggest trees of the world were represented "*at Home*" in my gallery, except the *Araucaria imbricata*, and I could find no description of this tree in any new books of Chilian travel; so, with the kind help of Sir T.F., a cabin was secured all to myself all the way to Valparaiso, and till we reached Bordeaux all was enjoyment. Then my nerves gave way again (if they were nerves!), and the torture has continued more or less ever since.

One night at the Valparaiso inn was enough for any one, though the place itself was full of life and work, its quays, iron landing-place, and docks, neat and efficient. Large ships could anchor close to the town, in the calm landlocked harbour. Its suburbs extended along the coast for some miles, Salto being the most attractive of them, as there was a valley full of the native palm (*Jubaea spectabilis*) which used to cover the country forty years ago; now, scarcely a hundred remain. They are strange, misshapen things, but seem quite in character with the rocky valley they grow in. The rail made a steep ascent to Quillota, where we came to the richest meadows and gardens, bordered by enormous hedges of the common blackberry, quite like the brambles of England, only gigantic; and rows of tall poplars. Huge bouquets and baskets of fruit were to be bought in quantities. The cherimoyers were famous there, and strawberries abounded, but they were all white or pale pink, none were red.

After leaving Quillota we had a still steeper ascent, among rocky hill-tops, with a sprinkling of cacti, puyas, and other purely native vegetation, till we arrived at the summit, then descended to the great wall of Santiago, with the city in the midst of it, and the snows of the Cordilleras beyond. Just as I was settling myself to dress and unpack comfortably in the hotel, there came a knock at my door, and I (thinking it was a jug of hot water) opened it in my dressing-gown, and let in Mr. Drummond Hay, our good consul, who had been sent to hunt till he found me; so I had to pack up again, and go off to Mr. and Mrs. Pakenham, the most hospitable of people, and they kept my room always ready for me, all the time I was in Chili, to return to at a moment's notice, and never put any one else in it.

I also made friends with Dr. Philippi, who lent me birds and wonderful nests from the museum. One little wren, "omnicolor," made an exquisitely finished nest, shaped like a small funnel; and I found another specimen of the same nest, with a loose, untidy dwelling on the top of it, built by a dowdy little brown bird, which had used the other as its pedestal. The *Acacia cavenia*, with terrible thorns pointing every way, was used to defend a curious nest by a delicate little bird called *Izuallaxis sordida*. It was entirely formed of these thorns, woven in and out, and the bird was very rightly called "the worker" by the natives. I tore my hands to pieces trying to get one of these nests, and had to give up the attempt; yet the tiny bird sat comfortably on a soft lining of hair and the sweet dry flowers of the tree, and seemed none the worse for weaving these terrible spines.

Of course the first thing I tried to get was the

great blue puya. I was told they were all out of flower; indeed some people declared they did not exist, because they had not seen them. At last an energetic English lady bribed a man to bring me one from the mountain. It was a very bad specimen, but I screamed with delight at it, and worked hard to get it done before it was quite faded, for it was past its prime. Then I drove out to Apoquindo, over a flat ten miles of uninteresting, cultivated country, with high mud banks or walls on each side of the road, which prevented one's seeing anything; but English weeds seemed to abound, and to grow with far greater luxuriance than they did at home.

My great object now was to find the blue puya, so I got a guide and a horse and started up to the mountains. We tied up the horses when it became too steep, and proceeded on foot right into the clouds; they were so thick that at one time I could not see a yard before me, but I would not give up, and was rewarded at last by the mists clearing, and behold, just over my head, a great group of the noble flowers, standing out like ghosts at first, then gradually coming out with their full beauty of colour and form in every stage of growth; while beyond them glittered a snow-peak far away, and I reached a new world of wonders, with blue sky overhead, and a mass of clouds like sheets of cotton-wool below me, hiding the valley I had left. Some of the groups had twenty-five flower-stalks rising from the mass of curling silvery leaves; about sixty branchlets were arranged spirally round the central stem, each a foot long, and covered with buds wrapped in flesh-coloured bracts; these open in successive circles, beginning at the base. I spent more than a fortnight working in comfort at Apoquindo, now and then mounting an old horse and riding into the mountains alone, tying the horse to some tree while I scrambled about after flowers.

But my chief object in coming to Chili was to paint the old *Araucaria imbricata*, known in England as the puzzle-monkey tree, rather unreasonably, as there are no monkeys in Chili to puzzle. Probably they crossed the Cordilleras in disgust at the general prickliness of all the plants there, especially the araucarias, and never came back again: there are plenty on the other side. It was not easy to make out how to reach those forests. People talked of difficulties, and even

dangers. They said I must sleep out, be eaten by pumas, or carried off by Indians, a noble race which had never yet been conquered by the white man. Others declared the trees no longer existed, having been all sawn up into sleepers for railways; but as usual I found when I got nearer the spot that all difficulties vanished.

After two days of railway through the long wide valley or plain of Santiago, with the snowy Cordilleras on one side and the high hills which go down to the sea on the other, I was received at the country-house of a half-English lady, whose husband was then going through all the troubles of canvassing the district, to be again elected a member of the Lower House.

Angol was the end of the railway there. The American Head of the Works kindly met me and allowed me to sleep in his house. I then found that the forests I was in search of belonged to two Irish gentlemen, one of whom came to fetch me the next morning at daybreak, and rode with me up a most picturesque and well-made road for some four or five hours, to the comfortable farmhouse where his sister and her family were living. From thence I could actually see the famous trees on the hill-tops, looking like pins loosely stuck into pincushions, as they stood out black against the sunset sky. Nothing could exceed the kind hospitality of my hostess, and no one could have wished for a more comfortable home. The house was very roomy, built as usual of one storey, with a verandah all round it, on a bare little knoll rising from green meadows and surrounded by hills, which were covered with trees resembling oak and beech. They grew separately, or in groups, so that the sun could peep through and sweeten the grass under them.

The ride up from dusty Angol had been very delightful. After mounting the first rocky ascent of 2000 feet, sprinkled over with puyas, cactuses, and other prickly plants, we left a glorious view of snowy volcanoes behind us, and entered on mixed forest and pasture scenery, passing stream after stream of clear running water, and more lovely flowers than I had seen in all the three months I had

Nest of the Trabajor Lesser Canestero (*Asthenes pyrrholeuca sordida* called *Izuallaxis sordida* by Marianne North) constructed from the thorns of the Espino-cavan (*Acacia cavenia*), Chili.

A Vision of Eden

passed in other parts of Chili. The embothrium or burning bush was in full beauty, growing in long sprays of six to eight feet high, quite covered with its vermilion flowers, which are formed like honeysuckles, but I saw no large trees of it, such as the one in my cousin's garden in Cornwall. Perhaps it enjoyed a new climate and soil, and throve in England as our common weeds flourish in Chili, where, on the Santiago plain, they had nearly driven all the natives out. The country showed instead an almost unbroken sheet of wild camomile, turnip, fennel, and different corn-flowers, far stronger than any we see in Europe. In the cracks near the streams were great masses of gunnera leaves (whose stalks are eaten like rhubarb), lovely ferns with pink stalks and young furry leaves, and among them, on the very edge of the streams, the *Ourisia coccinea* hanging its graceful stalks and scarlet bells over the water. The beech had its own pet parasite, a tiny mistletoe forming perfect balls of every shade of green and gold.

The first araucarias we reached were in a boggy valley, but they also grew to the very tops of the rocky hills, and seemed to drive all other trees away, covering many miles of hill and valley; but few specimens were to be found outside that forest. I saw none of the trees over one hundred feet in height or twenty in circumference, and, strange to say, they seemed all to be very old or very young. I saw none of the noble specimens of middle-age we have in English parks, with their lower branches resting on the ground. They did not become quite flat at the top, like those of Brazil, but were slightly domed like those in Queensland, and their shiny leaves glittered in the sunshine, while their trunks and branches were hung with white lichen, and the latter weighed down with cones as big as one's head. The smaller cones of the male-trees were shaking off clouds of

Chili

golden pollen, and were full of small grubs; these attracted flights of bronzy green parrakeets, which were very busy over them. Those birds are said to be so clever that they can find a soft place in the great shell of the cone when ripe, into which they get the point of their sharp beak, and fidget with it until the whole cone cracks and the nuts fall to the ground. It is a food they delight in. Men eat the nuts too, when properly cooked, like chestnuts. The most remarkable thing about the tree is its bark, which is a perfect child's puzzle of slabs of different sizes, with five or six distinct sides to each, all fitted together with the neatness of a honeycomb. I tried in vain to find some system on which it was arranged. We had the good fortune to see a group of guanacos feeding quietly under the old trees. They looked strange enough to be in character with them, having the body of a sheep and the head of a camel; and they let us come quite near.

The last place I stayed at in Chili was Las Salinas, a delightful house, hidden in a rich garden, ten minutes' walk from the sea-coast. Mr. and Mrs. J. were quite worthy of such a place. He was one of the best naturalists in the country. Under my window was a thick hedge of heliotrope more than a yard high. All the sweetest roses, with carnations and jasmine, grew there, as well as a huge *Magnolia grandiflora*, which shaded a large extent of ground and scented the air; but the wild walks along the shore were my delight. There were a good many fishermen's huts, and all the people seemed to employ themselves in collecting seaweed, of which a sort of isinglass is made and eaten. The cliffs were wreathed with mesembry-anthemums, calandrinias, cacti, puyas, fuchsias, oxalis, *Ephedra andina*, whose fruit is eaten and very sweet, and many fleshy plants whose names I did not know, reminding me of Africa.

I spent Christmas and New Year at Santiago, where Mrs. P., tried her best to make it look like home. I spent another night at the Salinas, and then steamed away in the *Mendoza*, touching at the nitre cities as we passed them, and landing on their miserable stony shores. I landed at Lima to recover my "nerves" and then went on to Panama, but I was too ill to go to Mexico and instead took my berth home. At Jamaica I landed early and drove out to see my old friend Mrs. C., who was then a widow, living three miles from Kingston,

The chief object of the artist's visit to Chili was to paint the famous Monkey Puzzle Tree or Chili Pine (*Araucaria araucana*) seen here under the Cordilleras of Chili with a group of Guanacos or Wild Llamas (*Lama guanicoe*) feeding among the trunks.

near her husband's grave. She screamed with delight at seeing me, and persuaded me to stay a month with her and have a real rest, as she was going up to a country-house on the hills for a change herself. I was easily persuaded, and she settled it all for me; quiet and much good feeding did me real good. Raymond Hall was one of the highest houses on the south side of the island, and the views were exquisite all round it.

1885. – Afterwards I returned straight to England, where it took me another year to finish and rearrange the gallery at Kew. Every painting had to be re-numbered, so as to keep the countries as much together as possible, the geographical distribution of the plants being the chief object I had in view in the collection.

A Vision of Eden

After that was finished, I tried to find a perfect home in the country with a ready-made old house and a garden to make after my own fashion, "far from the madding crowd" of callers and lawn-tennis.

1886. – I have found the exact place I wished for, and already my garden is becoming famous among those who love plants; and I hope it may serve to keep my enemies, the so-called "nerves," quiet for the few years which are left me to live. The recollections of my happy life will also be a help to my old age. No life is so charming as a country one in England, and no flowers are sweeter or more lovely than the primroses, cowslips, bluebells, and violets which grow in abundance all round me here.

Roblé (*Nothofagus obliqua*), a southern beech native to Chile and Argentina, together with its own specialised mistletoe (*Loranthus* sp.) parasite. Below are the leaves of *Gunnera chilensis*, on the left a bush of Winter's Bark (*Drimys winteri*), with behind bamboos (*Chusquea quila*), which are backed by other southern beeches (*Nothofagus* sp.), Chili.

CHAPTER XIV

Final Days

1886–90

(The following summary of Marianne North's twilight years was written by her sister Catherine Symonds.)

HERE HER RECORD ENDS. The long hard journeys were over, and, alas, with them was gone the greater portion of that indomitable strength which had seemed never to flag, which had carried her triumphantly through poisonous climates, never breaking down under incessant work, fatigue, bad food, and all those hardships which few women, travelling absolutely alone, would have dared to face. I have often thought how much her natural stately presence, and simple yet dignified manner helped her in facing all sorts and conditions of men in those long distant journeys. She inspired respect wherever she appeared, and good men everywhere were ready and eager to help her. Her work was always her first point: for that she travelled, not to pass the time, as so many mere globe-trotters do, in this age of easy locomotion. Her gallery at Kew is a monumental work: to finish it she fought bravely against increasing weakness; when it was done her strength was also gone, and the restful life she had dreamed of in her pretty Gloucestershire garden was not to be.

Yet one more year of hard work of a different kind remained to her. In the summer of 1886 she rented from General Hale, at Alderley in Gloucestershire, a charming old-fashioned gray stone house, with fields, orchards, and a garden, neglected hitherto, but exquisitely placed on the steep slope of one of those secluded valleys which

lie hidden away in the folds of the Cotswold Hills: an old-world region which took her fancy, and in which she determined to make for herself a paradise. That sleepy corner of West-Country England was soon astonished by her energy. Out of the dead level of the lawn-tennis ground she planned a terraced garden, sloping steeply to a pond and rockery which were to be stocked with rare plants from all corners of the globe. A little walled yard full of currant bushes she turned into a lovely rose-garden, sheltered by the old gray stable with its lichen-covered stone roof. The whole place had the rare charm, for an artist, of having been let alone for many years. Both trees and buildings were old, and all the trees had grown luxuriantly in that kindly West-Country air. When I came for the first time to stay with her in the Jubilee summer of 1887, I thought I had never seen so lovely a bit of English country as this which a chance had led her to.

The garden grew apace. Kew sent her all sorts of foreign rarities, a fine collection of cistus, and splendid todeas for the fern house. Her nieces at Davos collected Alpines for the rockery. All her florist-friends – Mr. Wilson, Canon Ellacombe, Miss Jekyll, and many others – sent generous contributions from their famous gardens: all were interested in her success. And *how* she worked! Every tiny plant, every bulb, was put in with her own hand or under her own eye, every label written by her, and entered in a book as the plants were fitted each into the nook which suited it best. All was order. In the mornings, long before six, before her men were out of bed, she was out upon the lawn with the garden-hose, patiently watering her fragile treasures; for that Jubilee summer was a terribly dry one, and very trying for those hundreds of young plants so recently imported into new surroundings. Every day fresh treasures came, in hampers and boxes of all shapes and sizes, from every quarter of the globe. Surely no garden on so small a scale ever had so much thought and loving care put into it. The good taste of her artist's mind helped her to utilise the things she found ready to her hand. An old sundial she had the luck to find in a neighbouring garden was set up on the lawn, a monument to the last survivor of the three little opossum-mice she had brought from Tasmania, whose virtues and voyages were recorded on a brass plate by J. A. Symonds, "Sir Henry" being buried underneath.

Many old friends came to stay with her that first summer and the next: the spare rooms were seldom empty. Professor Asa Gray and Mrs. Gray from Boston came with Sir Joseph Hooker, the Forrests from Western Australia, Mrs. Ross from Florence. But even then illness was creeping on her; she was never quite free from physical discomfort. The "enemy" of her last two voyages – those weary, constant noises – never ceased. In the quiet of her country garden, as in noisy London, the overtired brain still translated these into human voices, whose words were often taunts. Her deafness was of course responsible for this, and her brave common-sense recognised that the voices were delusions; still she suffered more than she owned, and dreaded being alone with those invisible but mocking foes.

In the autumn of 1888 a deep-seated disease of the liver, brought on originally, no doubt, by long exposure to all sorts of bad climatic influences, declared itself, and for many weeks she was as ill as it was possible to be. At Christmas we thought her dying; but her great natural strength, aided by medical skill and most careful nursing, enabled her to rally for a time, though her life had become one of constant suffering. The next summer she was better, and could even walk painfully about her garden and watch its progress. Strange to say, with the approach of active local disease, the unkind voices fled, and the relief of this was immense. But mortal illness was only arrested, not cured, and with her long habits of strength and self-reliance the life of invalidism and dependence on others became terribly trying. She died on Saturday, August 30th, 1890, and was buried in the quiet green churchyard at Alderley.

Her life there had only lasted five years, and of that period more than half had been shadowed by painful illness. But into the fifteen years immediately after her father's death had been compressed work sufficient for the lives of four ordinary women; and I have often wondered whether, if her strength had lasted another ten years, she could really have been content to sit down and wait for old age in the lovely green nest she had prepared for herself. Who shall say? She was a noble and courageous woman, whose like none of us shall ever look upon again.

The one strong and passionate feeling of her life had been her love for her father. When he was taken away she threw her whole heart into painting, and this gradually led her into those long toilsome journeys. They no doubt shortened her life; but length of days had never been expected or desired by her, and I think she was glad, when her self-appointed task was done, to follow him whom she had so faithfully loved. Her deafness, which had increased in the last years, really separated her painfully from the pleasure of daily intercourse with those around her. There is no sadder solitude than deafness. She painted as a clever child would, everything she thought beautiful in nature, and had scarcely ever any artistic teaching. In music it was different, and there, I think, her real genius lay. While her beautiful thrilling contralto voice, so absolutely true, lasted, she patiently submitted through long years to the drudgery of steady musical training under her mistress and great friend Madame Sainton Dolby. But that beautiful voice deserted her just when its cultivation had reached the highest point; and then painting, less cared for hitherto, was taken up to fill the void. She could never be idle.

When she once began to study plant-life, she read much and constantly about the trees and flowers she drew and cultivated, always with special regard to their out-door habits and characteristics. Her name has been given to five; four of which were first figured and introduced by her to European notice, viz. *Northea seychellana, Nepenthes northiana, Crinum northianum, Areca northiana, Kniphofia northiana.*

Marianne North 1830–90

"I am passing a very pleasant winter at home by the fire," wrote Marianne North to a friend from her London home in Victoria Street in 1879, "never going out of my flat except at night to meet agreeable people and hear music. I have just finished a large picture of Kinchinjunga to try its luck at the R.A. and I hope to paint a companion picture of a south country swamp to match it before April; and then when it is dark and yellow fog I scribble. I am writing 'recollections of a happy life' and putting all my journals and odds and ends of letters together. It is most amusing work even if it never comes to anything more."

Marianne was approaching fifty. She had travelled round the world alone, and had won a name for herself as a botanical artist, making it her mission in life to depict all the genera of flowering plants in their natural haunts, or as she put it in her characteristically simple vocabulary, "plants in their homes." Her travels were by no means over; she was yet to go round the world a second time; but she was at that age when the impermanence of life was beginning to perturb her. She had a longing to "leave something behind me which will add to the pleasure of others and not discredit my father's old name." Unmarried, she could not leave behind that natural testimony to her lineage which children would have afforded; without a confident religious faith the impulse for a tangible legacy, whether in paint or print, was more dear.

Two designs began to form themselves in her mind: one was the writing of her recollections, undoubtedly with the hope of publication, and the other was the building of a gallery at Kew to house her paintings. It was while she was waiting for a train on Shrewsbury station on 11th August 1879 that she wrote to her friend Sir Joseph Hooker, Director of the Royal Botanic Gardens at Kew, offering to pay for the building of a gallery and rest house in the gardens, where her pictures of plants could be instructively viewed near their live neighbours. "I am always for doing things off at once," she wrote, and she offered two thousand pounds for the construction, and proposed another friend, James Fergusson, as the architect. She was anxious that the gallery should be "a rest house for the tired visitors at the same time, with a cottage attached in which one of your married gardeners might live whose wife was capable of boiling a kettle and giving tea or coffee and biscuits (nothing else) in the gallery at a fair price." After much official correspondence her plan was approved, "all but the kettle boiling, which I must of course give up, though reluctantly." At that time, no refreshments were permitted in the Gardens. That her paintings should be popular and instructive was her great hope; she wanted to show people that cocoa did not come from the coconut; and she continued to add new paintings and to renew the decorations as long as her health permitted. "It was a great pleasure to see the thing looking so nice and full of people the other day," she wrote, "and gives me fresh courage to go on."

"It might be worth while to make the book in some sort a running commentary on the gallery," Sir Joseph Hooker suggested; and in book and paintings the same vivacity and enthusiasm carries the observer round the world and through a lifetime.

In the early years, Marianne's parents divided their time between the North family estate in Rougham, Norfolk; Hastings, where her father built a mansion in his Liberal constituency; and Gawthorpe Hall, Lancashire, home of her mother's first husband, Sir Robert Shuttleworth, whose premature death in a road accident left his wife with a delicate child to rear. Apart from her step-sister, Marianne had a brother, Charles, two years older, and a sister, Catherine, seven years younger. Throughout her childhood, it was on Marianne that her father chiefly doted: "She is the main link that binds me to life." When he found her "getting lazy and craving for excitement, wanting regular discipline and lots of air and exercise," he sent her to school in Norwich for eighteen months, and was gratified to find her returning "much softened in manner."

In 1847, when Marianne was sixteen, the family travelled abroad for three years on the Continent, seeing all too much of the revolutionary movements of the *annum mirabile*. Though she was bored by riding and dancing masters, she grew passionately fond of music. She had a fine singing voice.

When her mother, who had become increasingly an invalid, died in 1855, Marianne acquired a new role. Charles was away at college, and she was now mistress of the household. Her father had been re-elected to Parliament the previous year and decided to take a flat in London, "a most comfortable and neat little lodging," though the eighty steps deterred the girls from walking out, and being more in one another's company, they tended to be increasingly out of union.

It was about this time that Marianne began to devote herself to flower painting, which her father noted with some regret she "made a most exclusive business of." Each summer, after the Parliamentary session was over, father and daughters travelled on the Continent, usually to the spas of Switzerland and Austria, but also to Spain, where Marianne first tried landscape painting, and to Italy, Greece and the Bosphorus. All three carried the customary sketch book and diary, and Marianne's lively style led her friends to urge her to write up her travels. "I am told that I really ought to make some use of my travels and write

out a proper tidy journal from beginning to end," she wrote at the start of an account of their journey of 1861, illustrated with pen and ink cartoons and landscapes.

In 1863, when they were staying at Mürren, they found "two nice young Oxford lads" there, one of whom was writing home to his sister that "the young ladies draw very well and have good notions on art." By the next summer this nice Oxford lad, John Addington Symonds, the poet and critic, had asked Marianne's permission to court her sister, which he did most successfully in a romantic setting on a rustic bridge at Pontresina. Her father fully approved of the match, though he thought the young couple were "philandering rather beyond reason." Marianne meanwhile occupied herself in "sketching indefatigably."

Her sister's marriage in 1864 and her father's loss of his seat in Parliament the following year, brought Marianne into an even closer relationship with him. They walked together through Switzerland and the South Tirol, they wintered in Egypt and they trekked through Syria. Her father was unwell in Beyroot, and nervous of an expedition to Baalbek, but Marianne was "dying to be off with the tents, and thinking only of the bright side of pictures."

Painting still meant watercolour to her at that time, but the following year the Australian artist, Dowling, spent Christmas and New Year with them at Hastings Lodge and gave her lessons in oil painting, 'and I have never done anything else since, oil-painting being a vice like dram-drinking, almost impossible to leave off once it gets possession of one." The Hastings summer of 1868 was spent in "continual gardening," while her father began the weary work of canvassing, with his colleague Tom Brassey. A petition was lodged against them for bribery, and the case was not heard until the following April. Though it was dismissed and he took his seat, his spirit was all but broken. He was ill when they left for the Alps in August 1869, but to Marianne "all is simple enjoyment, and I say 'What's the odds, if you be pleased, Miss'." But in Austria his condition deteriorated and Marianne had to bring him home. They reached Hastings in October, and he died within a fortnight.

It is difficult to overestimate the severity of the blow to Marianne. There was the immediate shock: "It breaks one's heart to see her going about the house collecting little scraps that belonged to him, snuff boxes, spectacles, old memorandum books," wrote Catherine. "She is jealous that any other hands but hers touch them." And there was the long emptiness, never completely conquered.

There were practical problems too. Her father had been her chaperone in all her travels. That winter, she took her maid, Elizabeth, to Sicily, but found her companionship wearisome, and longed to be alone. "She looks well and handsome, very brown in the face," wrote John Symonds to his sister, "but she is not happy. There seems to be no future for her and she goes on painting in a ceaseless, mechanical sort of way as if to fill her time up."

But that spirit of restless animation which had always beguiled her father led her within two years to the first of those further excursions which were to shape the rest of her life. She had long wanted to visit the tropics to see and paint the vegetation. A chance acquaintance on their Nile cruise, a wealthy American widow, suggested that Marianne accompany her to the New World and be introduced to the best of Boston society. Marianne was excited both by the coast and country around West Manchester and by the people she met – the Adams of Quincy, Mrs. Agassiz, Charles Sergent and others. "Since I last wrote," she told a friend, "I have seen many wonders, and entirely lost my heart to – a poet! I mean dear old Longfellow! He is so very beautiful." The rich American widow was, however, abandoned when they reached Niagara, and Marianne travelled on alone from then, to meet, as she said, "less civilised and more interesting people."

Though she often yearned for solitude, no-one was more ready to enjoy the company of unpretentious people of all races and classes, whether met by accident or by design. She was sensitive to any form of hypocrisy, and always grudged time "wasted by cardleaving, dressing up and missionaries," while revelling in meetings with unorthodox or outrageous people. In Toronto she stayed with Judge Galt and his wife and nine children. Of the two eldest, one was "very pious, the other the reverse," and on the Sunday while the one urged the smaller sisters to sing hymns, the other led them in "Not for Joe." "Obedience was divided and the effect peculiar." From there she made an excursion to Pontiac to visit her father's old servant, John Loades, who had emigrated after their Hastings establishment was disbanded, and was now working on the railway. "John is only half the size he was," she wrote, "but he is very cheery. Betsy is a grand manager and makes even her own soap. John's wages are six shillings and soon will be eight shillings a day, and it is regular work for the whole winter, and he likes his boss, so he means to stay on and beat the ague if he can." On the return journey she had to spend a day at Chenoa, where she was much entertained by three "infant phenomena, full of airs and graces," who were going to give a concert in the evening with a "professor," whom she had no doubt was their father. "They told me there were six more quite as clever with mother at home, and when I took my departure they gave me two rosy apples and some peppermint lozenges as keepsakes." She always had a special *rapport* with children, respecting their honesty. In Kingston, Jamaica, the Monroe children "showed me little faults in my drawing and seemed full of intelligence."

Jamaica was a welcome retreat to "perfect quiet and incessant painting" after the excitement of Washington, where she had been taken to the White House by a Mr. Fish. Her shyness at having to "talk tête-à-tête with the prime minister in a brougham" left her when she was shown upstairs into a comfortable room where "Mrs. Grant's papa" was reading the newspaper, and the President's wife "fished up a big book of dried grasses to show me, and I put on my spectacles and knelt down on the floor to look at them and the great people found out I was not a duchess or worshipper of worth, and Mrs. Grant got quite at ease and told me she would never have let me come upstairs if she had known I was there but thought it was only Fish."

It was in Jamaica, living in the Garden House on two pounds of beef a week stewed up afresh with vegetables each day, that she began the collection of paintings of tropical vegetation which gave her a new purpose in life. Jamaica was followed by ten months in Brazil, during which she accompanied Mr. Gordon, manager of a mining company, and his daughter into the interior, with the colourful

"Baron" Antonio Marcus as guide. On this journey she learnt to expect any kind of accommodation, from a night in a barn, lost in the forest, or a grocer's shop, to a wealthy cigar manufacturer's mansion. She did not relish the novelty of deprivation, but almost any discomfort was acceptable if it took her into such rich and exotic vegetation. None of her later travels are recounted with so abundant a sense of excitement and enchantment. As they rode through the luxuriant forests she "grew wild with longing to dismount."

The excitement of Brazil was followed by a quieter year at home in 1874. She spent some weeks nursing her cousin Dudley who returned sick and wounded from the Ashanti War. She may have been ill herself; John and Catherine "thought her quite ill – we are anxious about her." When two acquaintances suggested she accompany them to Japan, all her friends were glad that she was not travelling alone; but she was thankful to lose her companions and spend long days in solitary sketching in the Yosemite, and contrived to sail from San Francisco by a later boat. In Japan the gardens, the courtesy of the people, the strangeness of their customs all enchanted her, but the visit was marred by rheumatic fever which left her with a fear of cold climates in which she always suffered.

After "a real good bed for once and a roaring English fire" in Singapore she went on to Borneo, where she was again dazzled by the vegetation, and was amused by the Rajah's native visitors, escaping however to a mountain farm when diplomatic visitors were due. It was on the steamer from Borneo to Java that she first met Dr. Burnell, the Sanscrit scholar with whom she maintained a lively correspondence until his early death in 1882. He was younger than she by ten years, though she may not have known it. She found his attitude to religion, society and literature engagingly unorthodox and admired his scholarship. "If you get ill," she wrote, "and worn out, send for me and I will come and nurse you and shock the world of old women as I did at Netley with Dudley – he the very most prejudiced and orthodox of unthinking soldiers and you a Bohemian like myself with thoughts on the most serious things which would perfectly dumbfound most of our best friends." Later she visited him at Tanjore. "I have found no

one in India who talks as you do," she wrote, "and very seldom meet anyone who takes interest in anything here. Mrs. Rivett Carnac is a real good solid kind of woman, but I think she would turn blue if she knew her guest was 'heathen'."

India and Ceylon were two countries to which she returned on her second world tour. The story of her first meeting with the famous photographer, Mrs. Julia Cameron, at Kalutara is well known: she "dressed me in flowing draperies of cashmere wool, let down my hair, and made me stand with spiky coconut branches running into my head and told me to look perfectly natural." On her second visit to Ceylon she was lodging in "delightful quiet quarters" from which she could see "the broad river running into the sea with rocks at its mouth half a mile off, and opposite me were all sorts of picturesque ferry boats crossing and recrossing and crammed with gaily coloured figures. Everywhere I looked I saw subjects for painting, and the sunset was as red as fire with the palm trees all black against it. I longed to stay quiet and enjoy it, but Mrs. Cameron found me out, and drove over the next day in the pouring rain and made me pack up and go back with her at once. My feet were in a pool of water all the way, and my poor wet sketch was sadly damaged in the transit, and I felt like a naughty child being taken back to school." In India she was plagued by the encumbrance of servants and their prevalent drunkenness. She escaped whenever she could from the Anglo-Indian society with its "tremendous dinners and picnic parties" and "croqueting-badminton young ladies," but the native art and architecture, religion and music, "exquisitely beautiful and barbarous," were a continual wonder to her. The breathtaking scenery of the Himalayas was worth all the cold to reach it: "The weather was finer every day, and clear till ten o'clock, and the great Kinchinjunga looked most grand as I saw it through the tangles in the forest. I got amongst giant rhododendron trees with pink trunks, not the common red variety but the sort Dr. Hooker found with pink and white and creamy bells."

In the summer of 1877 the Emperor and Empress of Brazil had visited her in her Victoria Street flat to see her paintings, an honour of which she was immensely proud. She had also been asked to lend her sketches for an exhibition in South

Kensington, and had spent much time selecting and cataloguing them before leaving for India. On her return two years later, finding the task of showing her Indian sketches privately to her many friends "very wearisome," she decided to take a gallery in Conduit Street for two months to exhibit them publicly. It was the success of this exhibition which prompted her to dream of a permanent home for her paintings, and to write to Sir Joseph Hooker offering to provide a gallery at Kew, and when she embarked on her next journey, to Australia and New Zealand, returning via Honolulu and a thirty-pound rail ticket across America, it was with all the advantages of official introductions to the most eminent botanists there. "People are extremely hospitable," she wrote, and the Governor gave me a free pass to all the railways. At Brisbane your first letter had a marvellous effect on old Mr. Hill at the Gardens, who actually offered me a room in his house and put a flourish in the local papers and I find everyone knows all about me wherever I go." She enjoyed Australia, though her journeys were by no means comfortable. In the Bunga mountains she "spent four days in tents in the greatest enjoyment," and on crossing a mountainside covered with *Zamia spectabilis* she wrote: "I went rather out of my mind at the sight of them." The ruthless killing of miles of "noble trees" near Fernshaw saddened her. She was acutely conscious of the need for conservation and especially for the conservation of trees. It was Charles Darwin who first suggested her Australian journey, and she was proud of his interest in her work. When he died, she said that he was the one person she would have liked to open her gallery. Sir Joseph Hooker was responsible for her programme. "I fear you will think me very idle for not painting more of the things on your list," she wrote, "but the Gardens do not tempt me – they are all so stiff and young."

After the warmth of Australia, her visit to New Zealand made her wretched with cold. But she had promised to visit her cousin there, and knew his mother would be disappointed if she did not. She was dismayed to find that her cousin's establishment was somewhat primitive, and involved "marching out of the doors through the cold to our breakfast in the kitchen house." Her fingers became too stiff for outdoor sketching. "I longed to be home without the trouble of going,"

she wrote, and she sat and painted from the window.

She reached Liverpool in June 1881 and spent a year preparing her gallery for the opening in the following summer. The arrangement of the pictures was to her own design; the painting of the door-posts and lintels, and the dado of specimens of woods beneath the pictures, were all her work. She took intense interest in every detail of the building, and of the catalogue compiled at her expense by W. B. Hemsley. But Africa, the fifth continent, was still unrepresented, and she set out in August 1882 to fill this gap. She was excited by the newness of the vegetation, yet beset by a curious indecision as to what to paint first, frustrated that she could not paint fast enough. In her letters from Africa there is a sense of growing tiredness and an increasing reluctance to be drawn into society. "I only go as far as my strength allows me: one can but die once, and how, it does not much matter to me. I want to growl at someone. I tramp morning and evening grumbling at my want of power to paint."

In May 1883 she returned home to get strong, see friends, and plan an extension to the gallery to take her African sketches, but within a few months set sail again for the Seychelles. The old enchantment returned at the sight of "these lovely islands. No paints can give half the colour of these seas." There she found the hitherto undiscovered female form of a rare orchid, and visited many of the islands in the government sailing boat.

Then things began to go wrong. The account in her autobiography is not explicit. The islands were subjected to quarantine because of suspected smallpox; drunken brawls at Christmas and New Year frightened and shocked her; she longed to be home, and though for a time she found peace, strangely enough, at a mission school where the company of children seems to have had its usual therapeutic effect on her – "They all seemed very happy and did not puzzle their brains with too much learning" – yet she was beset by delusions of persecution and danger which followed her to England. Restlessness drove her to Chile in August 1884, to find and paint the monkey puzzle trees and so fill another gap in her collection, but in spite of kindly hospitality she was upset by scenes of dirt and coarse living, bad taste and idleness. Travel had become a self-inflicted duty. She abandoned a

Marianne North 1830–90

plan to visit Mexico and returned to England for the last time in January 1885.

The purpose of her travels was after all as complete as it need be. She felt more and more "the need of perfect rest." She could not even stand the fatigue of going down to Kew, and began to look about for "some quiet cottage in a garden." The following year she found the house she was seeking, at Alderley in Gloucestershire, and devoted herself, as she had done at Hastings years before, to the making of a garden. She experienced anew the pleasure of farming and the society of country families, which must have reminded her of her childhood at Rougham. Something of her old enthusiasm and humour returned. "When are you coming to see the most perfect garden in England?" she wrote to a friend at Kew. "The next thing will be to have excursion trains run once a week to show you poor gardeners what the real thing is like." Her sister's three surviving daughters came to spend the Jubilee summer of 1887 with her; her favourite nephew, Fred, spent the vacations there with his law books. "Strength is gone and rest is coming," she wrote to Sir Joseph Hooker, "and come to some extent, enough to make life very enjoyable."

The enjoyment was short enough. In the winter of 1887–8 she fell ill. Janet Kay-Shuttleworth, her step-sister's daughter, was with her through the worst of it. Catherine, summoned from Davos for the third time that year, wrote that she was about as ill as it is possible to be. With remarkable tenacity she lived another two years, and was able to enjoy the company of the Galtons and other old friends in the summer. She died on 30th August 1890.

Of her two public memorials the gallery was standing, just as she had provided, at Kew; the *Recollections of a Happy Life* were still lying in manuscript, waiting for an editor to reduce them to proportions acceptable to a publisher. It fell to her sister to take on that task.

"Your aunt had a wonderful brain," wrote Catherine to her daughter, Madge Vaughan, more than a decade later. "It is good to be reminded of this now that her personality is already indistinct." But her personality shows through in her recollections as clearly as in the lines of her brush and the crisp colour of her paintings. It was, as an objective assessor had said of her book, "a bright sort of performance."

I should like to acknowledge with grateful thanks the permission kindly given me to quote from letters and papers in their possession by Mr. Roger North of Rougham, Norfolk; the Librarian of Somerville College, Oxford; the Librarian of Bristol University; the Chief Librarian and Archivist of the Royal Botanic Gardens, Kew; and the Director General of the British Library Reference Division.

Brenda E. Moon

Index

The following index lists species mentioned by their scientific (latinised) names in the main text (roman type) and illustration captions (*italic type*). The former where neccessary, by the staff of The Royal Botanic Gardens, Kew, to reflect their currently accepted scientific names.

Acacia cavenia 225, 226
Aciphylla 184
Acridocarpus natalitius 211
Acrocomia 74
Actinotus helianthi 164
Aethopyga ignicauda 146
Ailanthus altissima 145
Aloe bainessii 214; barbadensis 83
Alpinia zerumbet 50
Amherstia nobilis 20, 101
Ampelopsis 89
Angroecum eburneum 218
Anigozanthus flavida 171
Anopterus glandulosus 178
Anthocercis viscosa 168
Antirrhinum coulterianum 192
Aphyllon uniflora 43
Apodytes dimidiata 211
Aquilegia canadensis 43; chrysantha 192
Araucaria araucana 229; biduillii 158;
    braziliensis 64; cunninghamii 156;
    imbricata 225, 226, 229
Areca northiana 230
Arisaema triphyllum 40
Aristolochia brasiliensis 71
Artocarpus altilis 47, 92
Asthenes pyrrholeuca sordida 226
Banksia 172; grandis 172
Battus polydamus 54
Begonia 128
Bellendena montana 178
Bignonia venusta 48
Billardiera longiflora 178
Blighia sapida 48
Bombax malabaricum 116
Boöphone disticha 207
Boronia 164
Broughtonia sanguinea 56, 56
Brugmansia arborea 78, 110; sanguinea 56
Brunfelsia americana 56
Burtonia conferta 168
Byblis gigantea 176
Caladium 56; esculentum 48, 67, 83
Calamus tenuis 126
Callicore clymena 72
Callistemon citrinus 167; speciosus 171
Callitris tasmanica 177
Calochortus venustus 192
Calothamnus 63
Camellia 91
Carnegiea gigantea 194
Castilleja affinis 192
Catagramma 72
Cattleya 74
Cedrus deodara 130, 131
Celmisia coriacea 184
Centaurium venustum 192
Cephalandra palmata 207
Cerbera odollam 220
Cereus 74
Chamaebatia foliolosa 84
Chionochloa conspicua 184
Chirita urticaefolia 128
Chusquea quila 230
Cissus discolor 98; himalayana 130
Clerodendron 153; fallax 98; thomsonae 220
Cleyera fortunei 91
Cocos nucifera 115
Coelogyne asperata 103
Coleotype natalensis 211
Collinsia bicolor 192
Colubri 78

Comesperma volubile 172
Cordyline 184
Cornus nuttallii 189
Correa 164
Corynocarpus laevigata 186
Corypha umbraculifera 121
Crinum 52; augustum 156; northianum 6,
    230
Crocosmia aurea 211
Crotalaria 104
Cryptostemma calendulacea 168
Cucurbita digitata 194
Cyathodes glauca 178
Cymbidium finlaysonianum 154
Cyperus papyrus 24
Cypripedium acaule 43; calceolus 40
Dacrydium colensoi 184
Darlingtonia californica 26
Datura 104; metel 143
Daviesia 164
Delphinium variegatum 192
Dendrobium crumenatum 154; secundum 103
Desmodium oojeinense 126
Dicksonia antarctica 181
Dionaea muscipula 26
Disa graminifolia 201; grandiflora 200, 201;
    uniflora 201
Diuris 164
Doryphora sassafras 181
Dracaena draco 84
Dracophyllum traversii 183, 184
Drimys winteri 230
Drymophila cyanocarpa 178
Epacris 164
Ephedra andina 229
Eriostemon 164
Erythrina 68; variegata 143
Eucalyptus 163; cordata 178; diversicolor 174;
    globulus 178; macrocarpa 174, 176
Euphorbia pulcherrima 74, 126
Euterpe 47
Exocarpus cupressiformis 178
Ficus benjamina 94; bennettii 164; elastica
    118; moorii 164
Forsythia suspensa 91
Fouquiera 195; splendens 194, 196
Franciscea 64, 68
Fusanus spicatus 174
Gardenia 56
Gaylussacia baccata 40
Geranium maculatum 40
Gerardia 192
Gnaphalium 197
Grewia lasiocarpa 211
Gunnera chilensis 230
Hebe speciosa 186
Hedychium gardnerianum 146
Helichrysum 164
Hibiscus 211; liliiflorus 223
Hoheria lyallii 186
Hydnora africana 202
Hydrangea 104
Hyphaene thebaica 21
Hypoestes stellata 200
Hypolimnas misippus 207
Impatiens 104, 211
Ipomoea 211; bona nox 48; pes-caprae 213
Izuallaxis sordida 225, 226
Johnsonia lupulina 168
Jonesia asoka 126
Jubaea spectabilis 225
Juniperus virginiana 33

Kalmia latifolia 189
Kennedya 164; coccinea 168
Kingia australis 172
Kniphofia northiae 6; northiana 230; uvaria
    199
Koompassia excelsa 154
Lagerstroemia indica 220; speciosa 56, 149
Lama guanicoe 229
Lambertia 164
Layia platyglossa 192
Leschenaultia biloba 168
Lilium wallichianum 128
Liriodendron tulipifera 189
Lissanthe strigosa 178
Lithophragma heretophylla 84
Littonia modesta 211
Lodoicea maldavica 220
Loranthus 211, 230
Lyonia mariana 43
Macrosiphonia longiflora 68, 72
Macrozamia fraseri 172
Magnolia grandiflora 67, 202, 229
Mangifera indica 143
Melastoma 104
Melia azadirachta 138
Mellisuga minima 54
Mesembryanthemum 207
Metrosideros tomentosa 186; umbellata 184
Mimulus glutinosus 192
Montia perfoliata 84
Moringa oleifera 220
Mussaenda macrophylla 153
Myristica fragrans 54
Nelumbium 113; speciosum 108
Nelumbo nucifera 108
Nepenthes 94, 103; ampullaria 101; northiana
    6, 101, 230; rafflesiana 101
Northea seychellana 6, 222, 230
Nothofagus 230; cunninghamii 181; obliqua
    230
Nyctanthes arbor-tristis 143
Nypa fruticans 96
Ocyalas 66
Opuntia cochinellifera 83
Ornithoptera priamus, 153
Ornithorhynchus anatinus 163
Orthocarpus purpurascens 192
Ostinops 66
Othonna amplexicaulis 202
Ougenia dalbergioides 126
Ourisia coccinea 228
Paeonia suffruticosa 91
Panax crassifolium 183
Pandanus tectorius 125
Pandorea pandorana 167
Papilio polytes 92
Parochetus communis 104
Passiflora 156; alata 56; quadriglandulosa 56
Patersonia 164
Pavetta 211
Pentstemon azureus 192; centranthifolius 192
Petraea 172
Petrophila 168
Phacelia grandiflora 192; minor 192
Phascolartos cinereus 167
Philodendron 74; lundii 72
Phyllocladus asplenifolius 178, 181
Physalis 104
Pieris 92
Pimelea 164, 178; rosea 168
Phoenix dactylifera 21
Plascolartos cinereus 163

Ploceus capensis 209
Plumeria rubra 143
Poinciana regia 94
Polygonatum biflorum 40
Primula sinensis 91
Prostanthera 178; lasianthos 178, 181
Protea cynaroides 202, 204, 207
Pseudopanax crassifolium 184; gunnii 178
Psophodes crepitans 158
Puya alpestris 228
Ranunculas lyallii 184
Raoulia 186; eximia 186, 197
Rhododendron 91, 189; arboreum 134;
    griffithianum 134; nilagiricum 134;
    nudiflorum 43
Richardia albomaculata 199
Richea dracophylla 181
Robinia pseudoacacia 189
Roupellia grata 220
Roystonea regia 60, 64
Rubus australis 186
Russelia equisetiformis 220
Salisburia 89
Salvia columbriae 192
Saraca indica 143
Sarcodes sanguinea 84
Sargassum vulgare 44
Sarracenia flava 26; purpurea 26
Satyrium longicolle 207
Selasphosus rufus 189
Sequoiadendron giganteum 84, 86
Sequoia sempervirens 195
Smilacina racemosa 40
Solandra 63
Sollya fusiformis 168
Sphedamnocarpus pruriens 211
Stapelia variegata 207
Strelitzia alba 214; augusta 208, 213, 214
Strophanthus 220; gratus 60
Stylidium 168
Styphelia 164
Tacsonia thunbergii 48
Tecoma 220; mackenii 209
Tecomaria capensis 214
Tectona 107
Telopea truncata 178
Terpsiphone paradisi 116
Tetratheca 164; filiformis 168
Theobroma cacao 47
Thespesia populnea 123
Thunbergia coccinea 132, 137
Thysanotus 168
Todea africana 200; barbara 201
Tradescantia virginiana 192
Trichocereus chiloensis 228
Trichomanes reniforme 186
Trillium grandiflorum 192
Triteleia laxa 192
Trochilus polytmus 50
Vanda lowii 154
Vanilla phalaenopsis 218, 219
Vigna vexillata 211
Watsonia pyramidata 199
Wisteria chinensis 88
Wormia excelsa 94
Xanthorrhoea 172, 177
Xanthosia rotundifolia 168
Xylocarpus moluccensis 220
Xylomelum occidentale 167
Zygophyllum 207

T

Showing

Marianne North

illustrate

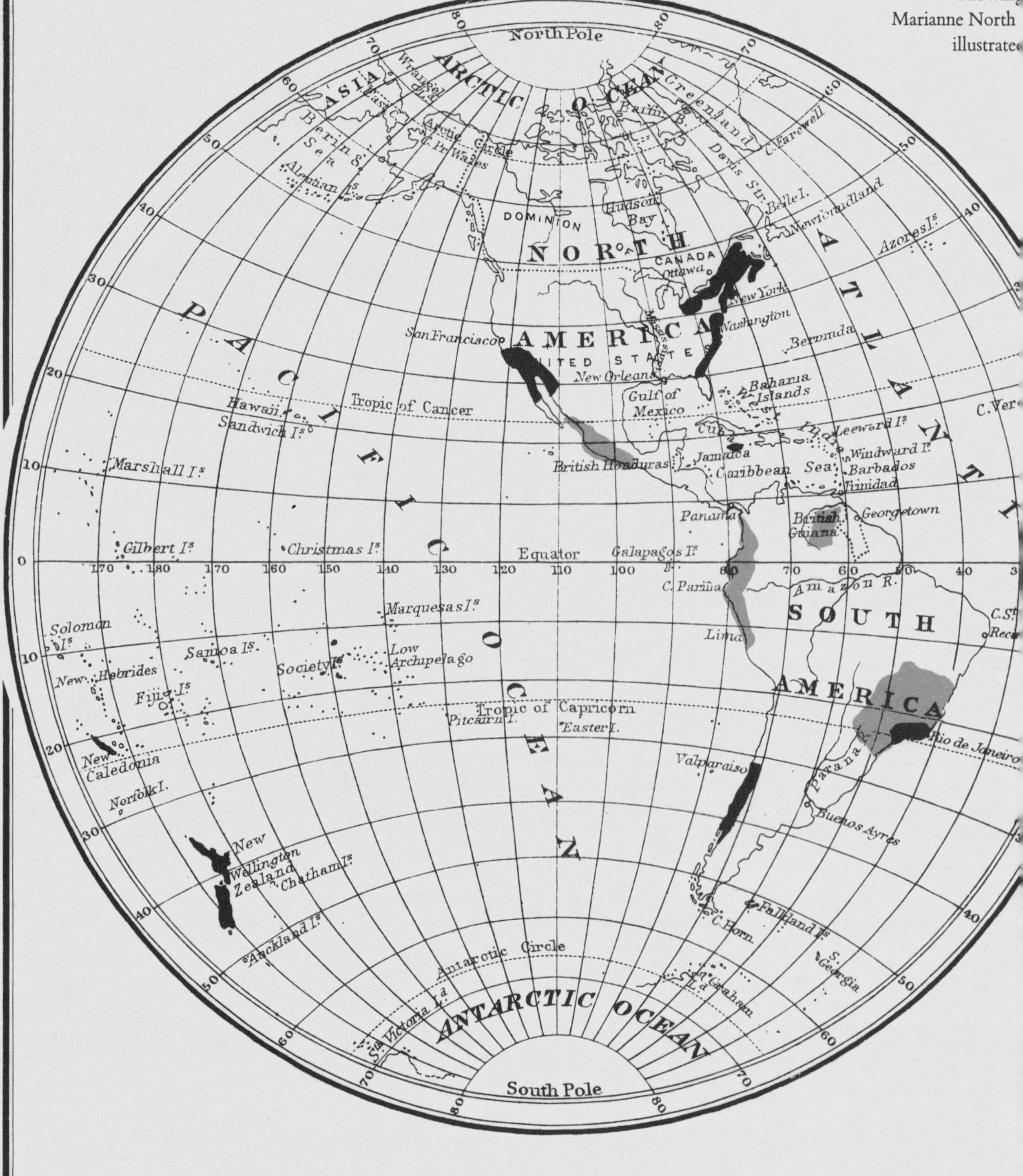